The Valley Section of Patrick Geddes

Walter Stephen

Patrick Geddes Memorial Trust
Publishers

ISBN 978-0-9555190-4-8

First published 2022

ISBN 978-0-9555190-4-8

The paper used is recyclable. It is made from low chlorine pulps produced in a low energy, low emission manner from renewable forests.

Typeset, printed and bound by Colorprinz, Penicuik, Scotland.

Contents

Illustrations – Colour Plates

Illustrations - Figures

Acknowledgements

My first action must be to record my gratitude to the Sir Patrick Geddes Memorial Trust for commissioning me to record and analyse Patrick Geddes's Valley Section, a fascinating task. In addition I was given considerable support by Trust members Sofia Leonard, Kenny Munro, Ben Tindall and by Murdo Macdonald, a former Trust member. The stained glass version of the Valley Section (and two other Geddes images) has been displayed and made accessible by the Scottish Historic Buildings Trust, whose support is gratefully acknowledged.

Even under Covid restrictions I was able to access our great national institutions, whose staff proved to be not only efficient but sympathetic. In particular I would like to thank Dr Anne Cameron (University of Strathclyde), Elaine Macgillivray (University of Edinburgh) and Lorna Black (National Library of Scotland). Kenneth Maclean in Perth was, as ever, a mine of information while Marion Geddes in the south of France reminded us that Geddes was no narrow nationalist.

The following illustrations are reproduced by kind permission:

Bulletin of Environmental Education – Fig 13.
Marion Geddes – Fig 9 and Fig 20.
Murdo Macdonald – Cover and Plate 1.
Juliet Robertson – Plate 8.
Kenneth Maclean and SAGT – Fig 22, Fig 23.
National Library of Scotland – Plate 5, Plate 7, Plate 10, Fig 14, Fig 15, Fig 16, Fig 17, Fig 18, Fig 19.
Patrick Geddes Memorial Trust – Plate 2, Fig 21.
Perth Museum and Gallery – Plate 3.
University of Edinburgh – Plate 5, Fig 1.
University of Strathclyde – Plate 4, Fig 2, Fig 3, Fig 4, Fig 5, Fig 6, Fig 7, Fig 10, Fig 11, Fig 12.
Yamaguchi Institute of Contemporary Arts – Plate 9.

Walter Stephen

CHAPTER 1 A VIGOROUS INSTITUTION

PHILIP BOARDMAN, GEDDES'S DISCIPLE AND BIOGRAPHER, summed him up as 'Biologist, Town Planner, Re-educator, Peace Warrior'. Paddy Kitchen's book is entitled *A Most Unsettling Person*. Lewis Mumford, probably the most influential American planner of the first half of the 20th century, when asked 'Who is Patrick Geddes?' answered thus: '...one might get the impression that Professor Geddes is a vigorous institution, rather than a man'. Later, the same man summed up Geddes, as follows:

> Turn by turn – and even simultaneously – Geddes was a botanist, economist, sociologist, producer of pageants, public lecturer, writer of verse, art critic, publisher, civic reformer, town planner, Victorian moralist, provocative agnostic, and academic revolutionary.

On the other hand, Alex Law, Professor of Sociology at the University of Abertay, Dundee, dismissed Geddes as 'a failed sociologist'.

Clearly this was a complex and interesting man, difficult to sum up briefly – and difficult to pin down. In his time Geddes was internationally influential. After his death he fell out of favour, but his ideas were rediscovered in the 1970s and he is now seen as a major figure in the environmental movement.

Patrick Geddes came of good Presbyterian stock. He did well at school and did 18 months in the bank, acquiring a banker's insouciant attitude to money. A spell of home education was to prove influential. He became a protégé of Huxley and Darwin at the Royal School of Mines, University College London and the Royal Botanic Garden at Kew. He was headhunted to set up the Scottish Zoological Station, the first in Britain. On a British Association collecting expedition to Mexico for Kew, Edinburgh and Huxley he became temporarily blind. Condemned to sit in darkness, Geddes used the experience positively, arranging his ideas and experiences into some kind of order.

In *An Eye for the Future*, an excellent television programme broadcast by BBC Scotland in 1970, there was a moving scene in which the late Leonard Maguire, a fine Scots actor and the very image of Geddes, recreated Geddes's breakthrough. We see the bandaged Maguire/Geddes emerging from the house mumbling and thinking aloud. Behind him the door has nine panes of glass in the upper half and 'in order to keep his place' Geddes feels for one pane and calls it something. He labels another, then another, then begins to think aloud about the relationships between one pane of glass and the next.

Geddes had discovered what he called a 'thinking machine' and what we might call a conceptual model or mental map, a deceptively-simple device for clarifying thought and explaining relationships. For the rest of his life his thinking machines were part of his stock in trade. It was another twenty years or so until the Valley Section began to come into being, when it was to become the most popular – and probably the easiest to understand – of his thinking machines.

Back in Edinburgh Geddes lectured in the School of Medicine and was appointed demonstrator at the Royal Botanic Garden. As well as his scientific work he involved himself in social and environmental issues. His eyesight might have made microscope work more difficult, but he certainly moved towards generalism. Yet he continued to write, often in collaboration, until 1931 *(Life: Outlines of General Biology)*.

In 1888 he was awarded the personal part-time Chair of Botany at Dundee. This was the perfect solution for Geddes. For a half-salary he would spend half the year at Dundee, have the security and prestige of the position and still be able to follow through his ever widening ideas in the rest of the year. Put quite simply, he was able to have his academic cake and eat it.

As a biologist his original work was overtaken, but his university teaching was not narrowly targeted at examinations, being more general, sometimes disturbingly so. For half the year he could develop his other interests, in three continents. Most observers look at society, say 'Something must be done' and then move on. Geddes was special in that problems inspired him into action and his enthusiasm was such that others had to follow.

Most of his enterprises were started in Edinburgh. Then he would generalise from his experience, elsewhere in Britain, in Ireland, in France, in Cyprus, in Palestine, in India. His big exhibitions - like his Cities Exhibition - won awards, toured the capitals of Europe and won over adherents in India. Retiring from Dundee in 1919 meant even more freedom.

During and after World War I he made three extended visits to India, preparing over fifty city plans and in 1919 being given the Chair of Civics and Sociology at Bombay (Mumbai).

Returning to Europe, he set up the Collège des Écossais in Montpellier, where he was able to dispense his ideas on a liberal education to an international clientele.

In 1932 he returned to London to accept his second offer of knighthood. Owing to King George's illness, he did not receive the accolade until 25 February. In London's toxic atmosphere he contracted an illness and returned to Montpellier, where he died on 17 April.

Thus he was Sir Patrick Geddes for only 52 days.

But Geddes's main legacy is a legacy of the mind. *Pelican in the Wilderness* (1956), *Silent Spring* (1962), *Blueprint for Survival* (1972), *Small is Beautiful* (1974) challenged the squandering of Earth's resources and the brutality of modern planning and architecture. Geddes was rediscovered, with his mantra 'Think Global, Act Local', in time for the Rio Earth Summit of 1992 and he is now the secular saint of the environmental movement.

His legacy is now a general commitment to careful and detailed survey of the problem, to the generation of an individual solution to an individual problem, to the genuine involvement of the people affected, in the planning and execution.

Pheroze R Bharucha, an Indian student who followed Geddes from Bombay to Montpellier, summed Geddes up thus:

> Assuredly there have been few like him – they hardly come once in a century. He just set you on fire with his love of this earth and with desire to cleanse it, to beautify and re-beautify it, to build and re-build it...

Geddes himself assessed his legacy when he said: 'I am the little boy who rings the bell and runs away'.

CHAPTER 2 A JOB AND A GOOD JOB TOO

22 OCTOBER 1851 WAS NOT A PLEASANT DAY for Acting Sergeant Major Geddes, recently returned with his wife and children, aged 12, 10 and 7, from foreign service in Corfu, Malta and Bermuda with the 42nd Regiment of Foot (Black Watch). On that day a Regimental Board at Aberdeen Barracks recommended his discharge 'as per Medical Officer's Certificate attached'. The Medical Report stated: 'The nature of this man's Disability is General Disability & Emaciation the result of long military service & Climate – has neither been induced nor increased by intemperance or other vice, he is "worn out" likely to be permanently disqualified'. Today we might think of Corfu, Malta and Bermuda as idyllic holiday destinations and Sergeant John Grant of Aberdeen, Pay Master's Clerk of the 42nd, was able to write to his girlfriend in 1844 from Malta:

> We have a fine airy situation in St Elmo and a delightful view comprising the City of Valetta, the Grand Harbour [and] Cattoneia and scarcely a day passes without a sail in sight, wafted along by the gentle breezes of the Mediterranean,

yet it was only after the Second World War that malaria was eliminated in these areas.

At that time every Highland regiment had a quota of six women per 100 men, who were employed on what we would consider support roles – washerwomen, cleaners, servants and the like. Janet Geddes was the regimental schoolmistress, with three areas of responsibility. There were the illiterate soldiers to be brought into line. Promotion for non-commissioned officers was dependent on the possession of Army Education Certificates and aspirants were brought up to scratch by Janet. And there were the children of the regiment to be taught. As a consequence, Janet had her own money.

At this stage it was intended that the family settle in Airdrie, where Mrs Geddes had a sister. This must have been a low point in the family fortunes. At the age of 43 Alexander Geddes, his wife and three young children were to be cast out from the security of modest seniority in the Army to the uncertainty of making a new beginning in an unfamiliar milieu.

On 25 November he was finally discharged at Chatham, being credited with **21 years and 338 days** of service. Later the same day, at an Examination of Invalid Soldiers at the Royal Hospital, Chelsea, he was awarded a pension of 2/- weekly (10p) based on service of **24 years 11 months.** (Note discrepancy of 3 years!)

My belief – which I cannot prove – is that, also on that day, Geddes met the Queen and Prince Albert. We know that the Royal couple habitually caused great offence to their eldest son and the Court by favouring their servants. In her *Leaves from the Journal of Our Life in the Highlands* of 1868 there are engravings of 'Ross the Piper, Brown and Grant' with 'MacDonald watching the deer, from a sketch by Prince Albert'. What could be more natural than personally briefing Geddes on a task so near to their hearts?

We next meet Alexander and his wife Janet Stivenson in Ballater, where they owned jointly two properties, an eight-roomed house and a six-roomed house with a flesher's shop, on a half-feu in a prime position in the centre of the village.
Alexander was now a civilian who was employed as a safe pair of hands to supervise the extensions and renovations under way at Balmoral, which Albert had bought in 1852. Albert was a real 'hands-on' person and dabbled in architecture. He and Geddes must have had interesting and meaningful conversations. The foundation stone of the 'new' Balmoral Castle was laid by the Queen on 28 September 1853. Geddes must have been there. Life must have seemed good for the Geddeses and it is not surprising that, at the age of forty, Janet gave birth to another son, Patrick (baptised as Peter).

The royal couple first stayed at the new Balmoral in September 1855 – but by that date Geddes had moved on. On 20 February 1855 a militia regiment, the Perthshire Rifles, was raised at Victoria Barracks in Perth. **Three days later** Geddes was commissioned into the Perthshire Rifles as Cornet or Ensign, **3 years and 3 months** after his medical discharge. He was promoted to Lieutenant in November and to Quartermaster the following year. This fellow who was 'worn out' in 1851 went on to serve for another 23 years and 37 days as a commissioned officer!

There is every sign that Victoria and Albert had found his services more than satisfactory. In 1857 his Service Record was amended to **increase his pension to 2/-6** on the authority of a letter from the War Office. And a month later another letter from the War Office docked the **three extra years** he had been awarded at Chelsea! (Covering their tracks?) In 1899 Alexander listed the particulars of his income. There was what he called his pension – his Black Watch service – and his 'Retired Pay' – his officer service. There was also a sum of £15 – an Annuity for Meritorious Service. What could have been more meritorious than 'special duties' carried out for the Royal family?

There is a strong smell of fish about this whole affair. How can we explain these strange goings-on? Geddes was headhunted. Coming on to the market at the right time he was a good, steady, reliable fellow and there was a job to be done. His pension was front-loaded. He had declined offers of a commission because he could not afford the costs. Here was a chance? A possibility? A promise? Of long-term security – if he came up to scratch. Which he clearly did.

Other biographers of Geddes have tended to be somewhat patronising about Patrick's father, dismissing him in a few lines as a lower middle-class authoritarian. But he was not just a modest military officer, he was a special military officer with a close – if short-lived – relationship with one of the Royal family's dearest projects.

WS Gilbert in the Gilbert and Sullivan comic opera *Trial by Jury*, summed up Geddes's situation very neatly. At the beginning of the trial, the Judge explains how he came to be in his lofty position, concluding:

JUDGE
It is patent to the mob,
That my being made a nob,
Was effected by a job.
ALL
And a good job too!

CHAPTER 3 THE SPECIAL ONE

WHEN QUEEN VICTORIA CAME TO PERTH, on 30 August 1864, three years after her husband's death, to unveil one of the first statues of Albert, it was Quartermaster Geddes of the Perthshire Rifles who was in charge of the proceedings, the protocol, the music, the guard of honour. Did she remember him? Did they exchange reminiscences? Or did the reality of the Queen's grief chill the whole occasion?

We cannot tell, but what we can be sure of is that, up at Mount Tabor, reminiscence would be flowing as Alexander looked back a few years to his association with Albert and the Queen, in the comparative informality of Deeside. It might have been 'When I discussed it with Prince Albert…' or 'The Queen was kind enough to say to me…'. Sitting round the breakfast table the family would have realized that Alexander was a Special One. And, knowing that the father was special must have given the rest of the family the thought that they were special, too.

Robert, the oldest son, married a general's daughter and ended up retiring early from being the head of the National Bank of Mexico, with a salary of £3,000 (worth £300,000 today). It is a family tradition that John McKail fled to New Zealand to avoid being pushed into the ministry. From a slow beginning he made a fortune in coffee and spices. The young Patrick turned up at the Royal School of Mines (now Imperial College), an ostensibly ill-qualified youth with an impenetrable accent from a distant and provincial Perth, yet became a protégé of Huxley and was respected by Darwin. The Royal School of Mines was full of bright young sparks from the public schools and 'good homes'. Yet, within a year, Geddes had risen above these others. His confidence in himself must have come, not only from the good Free Church background, but from his father's 'special relationship'.

What we know about Geddes's early days comes mainly from a series of articles written by him in the American *Survey Graphic* of 1925 (particularly *The Education of Two Boys*) and *Memories and Reflections*, written in Montpellier in 1928 for *The Young Barbarian*, the Perth Academy school magazine. Written for a purpose at a distance of seventy years can we expect objectivity and accuracy? Philip Mairet (*Pioneer of Sociology: The Life and Letters of Patrick Geddes,* 1957), who worked for Patrick during

World War I, also had a great deal of information from Arthur Geddes, but again at some distance in time from the actual events.

All the sources start with Captain Geddes in Perth – and, as we have seen, there was no Captain Geddes. No-one but myself, as far as I am aware, has troubled to investigate the matters described above and no-one else, again as far as I am aware, has troubled to note another, perhaps minor, reason for Geddes's self-confidence.

When the Geddes family moved from Ballater to Perth they first rented, then bought, a six-roomed cottage on the lower, wooded, slopes of Kinnoull Hill, across the Tay from the centre of the city. On his letters Alexander gave his address as Mount Tabor, a name whose significance seems to have been missed by commentators. Certainly, for Janet, Mount Tabor must have meant more than just a nice place to live.

Mount Tabor (*Jebel et Tur* – mountain of mountains – as the Arabs call it) is distinguished among the mountains of Palestine for its picturesque site, its graceful outline and for the remarkable vegetation which covers its rocky, calcareous side. There is a splendid view from its summit, which is traditionally the scene of Christ's Transfiguration.

This is where Jesus took Peter, James and John up the mountain:

> ...and was transfigured before them: and his face did shine with the sun...and behold a voice out of the cloud, which said, 'This is my beloved. Son, in whom I am well pleased, hear ye him'. (St Matt, 17, 1-9)

While I am not suggesting that Janet – or even Patrick – thought that her intelligent and questioning youngest child was a reincarnation of Christ, it would have been very easy to see the boy as a 'special one'. Already Alexander's career had been specially favoured, now his son was to grow up in a special place.

1843 was the year of the Disruption, when, on a point of principle, one third of the ministers of Scotland walked out of their manses and livings to found the Free Church of Scotland. In 1843 the Black Watch were in Malta, but the news spread like wildfire among Scots at home and abroad. In Perth Alexander became an influential Elder, especially in helping the congregation move to a new building in 1890. In their latter days the Sunday routine of Patrick's parents was to trudge down to morning service, then repair to the North Inch where they sat and ate their sandwiches (no cooking on the Sabbath) before returning to the church for the three o'clock service. Then came the uphill trudge home.

Norah, Patrick's daughter, gives us a snapshot reminiscent of Burns's *The Cottar's Saturday Night* when she recorded:

> Family prayers were said night and morning with readings from the scriptures. The maid was duly called in and when praying we knelt over the seat of our horsehair chairs.

The Free Kirk is often seen as the epitome of 'stern Presbyterianism' embodied in a plain lifestyle, no exuberant decoration, no highly emotional music, no drama and a nit-picking sabbatarianism – and no doubt this was true of many parishes. But Murdo Macdonald, in *Anarchy and the Free Church* points out that among the lay founders of the Free Kirk were some of Scotland's most significant thinkers of the time, scientists and artists. In 1905, when Geddes's Outlook Tower was in financial trouble, it was a Free Kirk minister who gave the keystone lecture and when a management committee (with 76 members!) was formed it had a significant element of Free Kirk ministers and laymen who had graduated from the Free Church College.

While Geddes fell away from adherence to any church he made it clear that he owed a debt to the Free Church, saying that it was the organisation of which he was 'proudest of all' to have belonged to. Significantly, in later life, he was to say that his favourite Biblical text came from Nehemiah, the prophet responsible for rebuilding the walls of Jerusalem, despite the opposition of the king and rulers. Nehemiah anticipated Geddes's planning model; he carried out a survey, made a plan and carried it out with the community's involvement. And there followed a ceremonial dedication.

As Nehemiah said (Neh. IV. 4-6):

> So built we the wall; and all the wall was joined together unto the half thereof: <u>for the people had a mind to work.</u>

(Author's emphasis).

Not a bad recipe for a life's work!

CHAPTER 4 PG AND SCHOOLING

ONE DOES NOT KNOW HOW PEOPLE TALK these days about the curriculum, but not so long ago we used to talk about the formal and informal curriculum, behind and above which lurked the hidden curriculum. For Geddes the Free Church was a major factor providing attitudes and a framework for later life.

As for the informal curriculum, Geddes did not come from a deprived background, indeed, it could be said that his background was privileged. His father was decent and respectable, comfortably off and decently employed in a post which left him plenty of time in which to work with his young son. Patrick grew up in a caring household of three women, a competent mother, an unmarried elder sister who was devoted to him and a mature live-in servant who was, in effect, one of the family.

Patrick Geddes grew up in a garden. A garden on the slopes of Mount Tabor, itself a 'foothill' of Kinnoull Hill. A garden full of useful produce and abundant with the flowers his mother loved. The little Geddes followed his father around with his wheelbarrow while 'the kind father' used the planting of potatoes to give him his first lessons in mathematics. On Sunday afternoons no work could be done, but the whole family would walk around examining every plant in detail and planning the week ahead.

The wooded Kinnoull Hill was a marvellously rich adventure playground and a viewpoint ('a nobly wooded hill, with...a long ridge of noble precipice') which stirred the imagination. To the north were the Grampians, and in front of them was some of the most fertile land in Scotland. Great rivers came together to make Perth a port from Roman times. As the poet McGonagall almost wrote:

> The Tay, the Tay, the silvery Tay,
> It flows from Perth to Dundee every day...

And did write:

> No other river in the world has got scenery more fine,
> Only I am told the beautiful Rhine.

The young Geddes collected interesting things he found in his explorations and 'a tendency to mischief' was cured by his father converting a lean-to shed as a laboratory, with a carpenter's bench as well. (All his life Geddes was to preach the educational doctrine of 'Head, Heart and Hand' and to commend the making of a box as the basic educational experience.)

Father and son went on long walking tours in the Eastern Highlands, storing up observations which would re-emerge in later years. Alexander Geddes had a brother who was Postmaster and General Merchant in Braemar, 58 miles north of Perth. Via the Devil's Elbow and The Cairnwell they walked there and back (as did Jessie, his sister, in 1864 – alone).

In later years these various juvenile observations were to be systematised and come together in the Valley Section model.

In 1871, the year in which Geddes left school, Carl Fleckstein, teacher of Modern Languages at Perth Academy, proposed Geddes's name to the Perthshire Society for Natural Science, and a year later Geddes was voted on to the Library Committee. This was no collection of hairy-handed mechanicals but a serious society with a fine library (by coincidence, perhaps, it bought in the same year, *Descent of Man* by Charles Darwin) and at least two members of distinction.

James Geikie was the District Surveyor of the Geological Survey of Scotland who went on to be Director-General of the Survey and then Murchison Professor of Geology at Edinburgh and a world authority on *The Great Ice Age and its Relation to the Antiquity of Man*. Dr Francis Buchanan White, like Darwin, was a gentleman-naturalist. He took great pleasure in stopping people in the High Street of Perth to show them his latest letter from the great man. He had several important papers published and did apply, without success, for several Chairs. As 'the Macbeth of Scottish natural history' he was the driving force behind the PSNS, going on to become the founder/editor of the *Scottish Naturalist* and a leader in the movement to set up Naturalist Societies in Scotland.

Geddes played a full part as a member of the PSNS, attending every lecture and networking to good effect, so that, in later years he would return to Perth as an expert and for financial support.

After eighteen months 'in the bank', acquiring that insouciance about money that bankers have, Geddes was allowed a period of 'home studies' – quite remarkable for strict Victorian parents. He had his workshop/laboratory/museum and had lessons on cabinet-making in the mornings and attended the School of Art in the afternoons. He was encouraged to read voraciously in Perth's seven circulating libraries and six good, serious libraries. Darwin's books were in circulation and Geddes, on behalf of the Library Committee of the PSNS, would be able to order whatever he needed.

As for his formal education we turn first to John MacKail Geddes, Patrick's brother, his elder by eight years, who wrote from New Zealand when Patrick was aged nine:

> I am happy that you are always keeping up your place as a gentleman and a scholar. You will soon be fit to come to a decision what you will want to be. Whether philosopher, gardener, astronomer or militiaman.

A year later his brother wrote again:

> Last month being the time of the Perth examination Pat would be very busy and I hope to hear that he has got several prizes,

and,

> I was very glad to hear that you have got three prizes. You richly deserve them. Would you tell me what they are for… Father will be prouder about them than you are yourself.

At Perth Academy he went on to achieve University entrance, without trying too hard. He had a good grounding in English, French, Chemistry and – especially – Maths, where he was very well taught, partly in the outdoors. He 'soon tired of Latin'. In the Perth Museum and Art Gallery are his *Fair Book* (of Mathematics), *Mensuration of Distances* and a Merit Certificate of the Scotch Education Department of 1870 in English, Geography and History.

During his Home Studies he had formed a 'burning desire' to study under Huxley. He attended Edinburgh University for a week, then rejected what it had to offer and moved to the Royal School of Mines (from 1907 Imperial College). There he had to do a first year probationary course. He bet:

> ...some of the engineering students that he could pass the elementary examination in mining after only one week's study. He crammed the requisite text books for seven days and duly passed, eventually receiving a certificate which entitled him to become a sub-inspector of mines – although by then he had forgotten virtually all his ill-digested mining knowledge!

This certificate could be said to be the end of his formal education - and of this chapter! However, his education continued, without formal recognition. Having impressed Huxley by picking up an error in a draft paper, he became 'something of a real assistant' to Huxley, was put on to his own research, was found a place as a demonstrator at Kew and was put up for the Sharpe Scholarship at University College London. 'Moonlighting' at the Birkbeck Literary and Scientific Institution he tutored Annie Besant – seven years his senior – the social reformer. They were to co-operate many years later, in India, where she had become President of the 32nd Indian National Congress.

In 1878 Huxley sent Geddes off to the marine biological station at Roscoff, in Brittany. He spent the winter at the Sorbonne, making contacts that would last a lifetime and then moved to Naples, where Anton Döhrn had set up the first zoological station and aquarium in Europe. Here Geddes had research space, facilities and a supply of fresh specimens. Huxley had been elected Lord Rector of Aberdeen University and James Cossar Ewart (who had shared laboratory space at UCL with Geddes) was made Professor of Natural History there. Now Geddes was appointed 'director' of the Scottish Zoological Station at Cowie, just outside Stonehaven. It was only seasonal – but it was the first in the country!

The next step was the award of a £50 grant from the British Association for research in paleontology and zoology in Mexico, sending specimens back to Kew, Huxley and Edinburgh. He had a spell of temporary blindness and it was during this that he began to systematise his ideas into what he was to call 'thinking machines' – the most famous of which was to be the Valley Section.

He came back to Edinburgh University to a lecturer/demonstrator post in the medical faculty and Royal Botanic Garden. He taught, wrote and published but also began diversifying his interests into social and environmental issues and projects. He applied for several chairs, without success, but in 1888 J Martin White, a boyhood friend, endowed him with the Martin White Chair of Botany at University College, Dundee. Geddes was paid half the salary of a full-time professor but taught for only part of the year with an assistant paid to do the donkey work. As Philip Boardman said:

> Geddes could earn his academic cake in Dundee, so to speak, and eat it for three-quarters of the year anywhere else he wished.

We might say his formal education was now complete. He had a measure of financial security and, most importantly, the prestige of a professorial title. From that basis he proceeded to roam around, in Europe, North America and Asia, exploring the interests wider than biology that were now engrossing him.

CHAPTER 5 *THE BEST VALLEY SECTION*

IN HIS 1905 LECTURE ABOUT THE OUTLOOK TOWER John Kelman described a certain stained glass window as 'beautiful and impressive', with the motto: 'Microcosmos naturae, sedes hominum, theatrum historiae, eutopia futuris'. He said it was a little landscape and went on to analyse its symbolism. Here was the education of the eye. The window, once in the Outlook Tower, is the property of the Sir Patrick Geddes Memorial Trust and is on loan to the Scottish Historic Buildings Trust. It is on display in Riddle's Court, which was one of Geddes's self-governing student residences.

In my view it is the best Valley Section because of its having been the first working out of Geddes's concept for didactic purposes, not lost but happily settled in an eminently suitable location. As you can see, it is big, colourful and crowded with symbolism which bears close inspection. Like the stained glass in a medieval cathedral the story is there for all to receive through the eye – except that this is the only Valley Section to have the mottoes along the bottom. (Plate 1)

What do we see when we look over Murdo Macdonald's shoulder? Basically, the long profile of a river valley-side, from the high ground on the left down to the sea, and peopled by a variety of 'occupational types'. At the bottom left we see the entrance to a mine and a quarry with a protective fence around it. Here labours the Miner. The high ground is the province of the Woodman. The Hunter patrols its lower margins. Sheep, cattle and patches of beige illustrate the realms of the Shepherd and the Peasant, with his modest cottage. The Farmer works the big flat ploughed fields. The harbour and the sea provide a living for the Fisher. The ship in the bay is no fishing boat but is probably from Russia or the Baltic, loading a cargo of salt herring for the winter.

Geddes backs up his visual message with Latin phrases along the bottom: *Microcosmos Naturae, Sedes Hominum, Theatrum Historiae, Eutopia Futuris*. He is suggesting, nay, directing, 'that we look again at the valley from a set of contrasting and yet illuminating viewpoints'. For Geddes the biologist the valley was above all a 'microcosm of nature'. To understand the biology of our planet, we start in the microcosm of our own locality. For Geddes he was to move from the Tay on its course from the Moor of Rannoch via Perth to Dundee, to the Rhine, the Danube, the Ganges flowing from the mountains of the Himalayas to Calcutta.

But the same river valley defines the *sedes hominum* – the abode of human beings, where people are born, live their lives, have their memories, and die. Nature is still there but is developed into a model of society, its culture and institutions.

With the *Theatrum Historiae*, the theatre of history, we are moved from the environmental emphasis of the valley to the cultural. Not only through the realities of archaeology and history, the creative reimaginings of architects, artists, poets and novelists but through the understanding of history and all its related disciplines we can look in a clear-sighted way to the future.

We have been moving from left to right and now we come to Eutopia Futuris, the 'eutopia' or 'good place' of the future. Sir Thomas More postulated an 'utopia' with a 'u' – an idealized 'no place'. 'Eutopia' with an 'eu' is a good place that Geddes believed could be achieved through local and international co-operation.

Yet Geddes's didactic symbolism does not stop with these words, but goes on to the skies above the valley. Top left are two game birds linked to the idea of hunting, with a suggestion of a more primitive, immediately nature-dependent, way of life. Above the theatre of history are two fighting birds of prey (eagles?), portraying, perhaps, our present, institutionalised, historically-conscious view of ourselves. Geddes was no soppy, romantic fool, he offers us an alternative – the 'kakatopia' or bad place of the future. Well over a century ago he knew we had a choice between ruining the planet and enabling it to sustain itself.

Remember these words of his:

> ...this is a green world, with animals comparatively few and small, and all dependent upon the leaves. By leaves we live.

His clear sight is to his credit, but at the same time we are humbled by realising how long his environmental message has been ignored - and that it may be already too late to save the planet.

But above Eutopia we can just make out three doves in harmony, symbolising a transition from a crudely mechanised society based on unfettered competition, to Geddes's dream of a society which made use of sophisticated, environmentally-friendly technologies, based on local and international co-operation. Geddes loved triads – Work, Place, Folk, an education of Head, Heart and Hand. His planning model was Sympathy, Synthesis and Synergy.

His three doves appear in many Geddes contexts – for example, when Castle Wynd South was renamed. (Plate 2).

In Paris, in 1878, Geddes met the French geographers Le Play, Élisée Reclus and Edmond Demolins and picked up Le Play's *Lieu, Travaille, Famille*. Significantly, Geddes transformed this into Place, Work, Folk. Even before meeting Dewey in America in 1898 Geddes was advocating an education of Head, Heart and Hand, claiming that he had had this and providing it for his own children.

In the Valley Section window the doves suggest that the Eutopia of the future will be brought about by the Geddes planning model of Sympathy, Synthesis and Synergy. Sympathy was studying and surveying the problem, gathering the information with the help of the people. Synthesis was putting it all together to evolve a plan, while Synergy was all working together to put the plan into action.

Thus, as Murdo Macdonald says on his essay on *Patrick Geddes: Environment and Culture*:

> In one image, the stained-glass version of the valley section diagram, Geddes provides the basis of a philosophy for thinking about human ecology, not only by drawing attention to the necessity of exploring any environment in terms of its natural characteristics, and the way in which those characteristics relate to the folk who live there and the work done, but also by specifying the psychological attitude necessary for any such analysis, that is to say an emotional, intellectual and co-operative engagement with that place, those folk and their work, to use one of Geddes's favourite combinations of terms.

CHAPTER 6 EARLY INFLUENCES

THE STORY BEGINS WITH THE PUBLICATION in 1658 of *Orbis Sensualium Pictus* by John Amos Comenius (1592-1670). The first widely used children's textbook with pictures (and 150 chapters!) it was in German and Latin (the international language). It quickly spread to other countries - the first English edition was published in 1659 and a four-language edition in 1685. For centuries it became the defining children's textbook, being continually in print until 1780, with upgrading of pictures and text content. 'For some time it was the most popular textbook in Europe and deservedly so'.
And we can still trace his influence in North America!

Comenius was an original thinker and innovator, most of his ideas being startlingly modern. Pictorial textbooks in native languages. Teaching from simple to more comprehensive projects. Lifelong learning. Equal opportunities for women and children. Universal and practical instruction.

Comenius's ideas have been summarised thus;

(1) learning foreign languages through the vernacular;
(2) obtaining ideas through objects rather than words;
(3) starting with objects most familiar to the child to introduce him to both the new language and the more remote world of objects;
(4) giving the child a comprehensive knowledge of his environment, physical and social, as well as instruction in religious, moral, and classical subjects;
(5) making this acquisition of a compendium of knowledge a pleasure rather than a task; and
(6) making instruction universal.

In his *Orbis Pictus* Comenius developed for the modern era the notion of visual experience as integral to verbal explanation. In Kelman's 1905 lecture on the Outlook Tower we can see where he and Geddes got their total emphasis on observation and survey in the field, not in the study.

> For what more can naturalist or geographer claim to possess than the habit of observing and thinking for himself and at his best, without books or helps, in presence of the facts, and in the open air?

Thus spake Geddes in *Nature Study and Geographical Education*, in the Scottish Geographical Magazine of 1902. And in Comenius's diagrams – not just pretty pictures – we can surely see the germ of Geddes's 'thinking machines'.

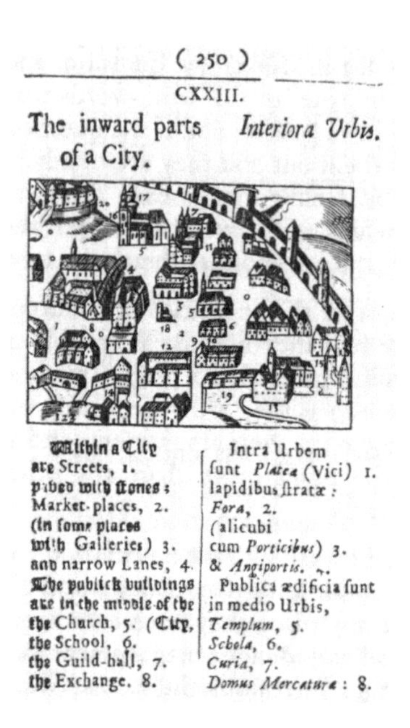

(250)

CXXIII.

The inward parts of a City. *Interiora Urbis.*

Within a City are Streets, 1. paved with stones; Market-places, 2. (in some places with Galleries) 3. and narrow Lanes, 4. The publick buildings are in the middle of the City, the Church, 5. the School, 6. the Guild-hall, 7. the Exchange. 8.

Intra Urbem sunt *Plateæ* (Vici) 1. lapidibus stratæ: *Fora*, 2. (alicubi cum *Porticibus*) 3. & *Angiportis*. 4. Publica ædificia sunt in medio Urbis, *Templum*, 5. *Schola*, 6. *Curia*, 7. *Domus Mercaturæ*: 8.

FIG 1
The inward parts of a City (*Orbis Sensualium Pictus*)

Lord Kames (1696-1782) was one of the 'founding fathers' of the Scottish Enlightenment. David Hume called him 'the best friend I ever possessed' with 'an iron mind in an iron body'. Not only was he a brilliant advocate of 'tireless energy', but also, according to Hume 'the most arrogant man in the world'. He had a great deal to be arrogant about, not least being his role in the case of Joseph Knight, an African-born slave brought to Scotland in 1769, when Kames stated – 'We sit here to enforce right, not to enforce wrong' – and Knight was set free. At the age of seventy he took over the management of his wife's estate of Blair Drummond and began the reclamation of the carseland west of Stirling.

In a collection of essays in 1758 – *Historical Law Tracts* – Kames came up with a momentous and far-reaching concept, the organisation of the history of the human community into four distinct stages. 'Hunting and fishing were the original occupations of man', and tended to be, by their nature, solitary. Then came the second stage, the pastoral-nomadic stage, when men had followed the herds and domesticated them. 'The shepherd life promotes larger societies' of clans and tribes.

The third stage is that of agriculture. 'Cultivating the fields is by necessity a communal enterprise'. Not just the annual harvest but farmer/craftsman, landlord/tenant, master/slave. New forms of co-operation, competing interests and conflict emerge. The fourth stage is 'commercial society', born of the buying and selling of goods and services.

William Robertson took Kames's four-stage theory and applied it to the history of Europe since the fall of Rome, creating the modern study of history and a genre - 'the story of civilisation' - which circulated for 250 years.

Geddes took Kames's 'model', refined it and made it visual. Kames identified a progression, the Valley Section is also a progression, from high to low, but with occupations in more than the four stages. (Seven in the original version, eleven in one of the 1920s versions). On the left, high up, we have the solitary Hunter, with the Woodman (hunting in the forest) and the Miner. Next we have the Shepherd, his sheep and cattle. Stage 2. Then we have the simple cottage and fields of Stage 3, but Geddes divides it by having a Peasant (in some versions with a spade) and a Farmer (with a plough). Stage 4 in Geddes's Section is a little town, a harbour and a ship offshore. Geddes said it was the domain of the Fisher, but it is clearly more than that. It is commercial society - Kames's fourth stage.

High up on the walls of the Perthshire Natural History Museum (founded in 1883) were four geological diagrams. No. 2 was

'Section through the Tay Valley at Woody Island' which was copied and is now in store in Perth Museum and Art Gallery.

Plate 3 shows Lower Old Red Sandstone overlaid by Boulder Clay and Estuary Clay, the latter with much peat and overlaid by river gravels in three terraces. Place names and altitudes are given. A Basalt Dyke cuts through the sandstone, predating the Boulder Clay above.

In 1883 Geddes was in Edinburgh but was by then a Corresponding Member of the Perthshire Society of Natural History. He returned to Perth for a *Conversazione* in the Museum, opening the Saturday session with an address on The Study of Biology. Geddes must have seen these four diagrams, with recognition. No. 2 was a section across a valley – a very common illustration – a valley section, if you like. But it is not a Valley Section. Geddes knew his local area, the shepherds, the farmers, the foresters, the salmon fishers, and it is probable that, from that day, he began to formalise their situations in the landscape. In the process he set aside the cross section of the valley and instead promulgated a longitudinal section from the heights to the sea.

It is generally agreed that an important influence on Geddes was *Histoire d'Un Ruisseau* by Élisée Reclus (1830-1905) geographer and lifelong friend of Geddes. Reclus was a keen Republican and pacifist who approved of theft, as a revolutionary act. For him, all property was theft. Reclus's output was phenomenal. He travelled widely and from 1893 lectured at Geddes's Summer Meetings Eg on *The Evolution of Rivers*. Under the title *A Great Geographer: Élisée Reclus, 1830-1905* Professor Patrick Geddes wrote his obituary in fifteen pages for the Scottish Geographical Magazine. Reclus had a posthumous moment of celebrity when his views on the suitability of rivers and river valleys as international boundaries were considered (and ignored!) at the peace conferences which led to the dismemberment of Hungary after the 1914-18 war.

Histoire d'Un Ruisseau, first published in 1869, is essentially a description of a river from source to mouth, in 300 pages and 20 chapters. The chapter titles give a fair impression of the style of the book as follows:

> The Source – Water in the Desert – The Mountain Torrent – The Cave – The Sinkhole – The Ravine – The Valley Springs – Rapids and Cascades – Sinuosities and Eddies – Flooding – The Banks and Islands – The Promenade – Bathing – Fishing – Irrigation – The Mill and the Factory – The Timber Train and Ship – Water in the City – The Estuary – The Water Cycle.

Clearly Reclus's plan is to describe the course of the river – physical geography, if you like – and add to it human activities – work and leisure. It could be considered as a model for a river basin. It is not a Valley Section, but it is on its way to being one.

The American William Morris Davis (1850-1934) began as a meteorologist before moving into geology and physical geography. In an article of 1889 - *The Rivers and Valleys of Pennsylvania* – he put forward the 'geographical cycle', the cycle of erosion which suggests that (larger) rivers have three main stages of development, usually considered as youthful, mature and old age, each with its distinct landform and character.

While Davis's cycle is purely physical it is easy to see how Geddes would seize on it to show the Woodman and Hunter in the youthful valleys of the mountains, the Farmer in the good soils of the mature river and the Fisher down where the 'old' river merges into the sea.

Lewis Mumford, the great American planner, spent five days with Geddes in the Outlook Tower in September 1925, when Geddes was 'making efforts to tidy up the middens of notes and manuscripts' in the Tower. Later Mumford was to recall that, 'in the midst of my packing... tidying up after Geddes was like putting the contents of Vesuvius back into the crater after an eruption'.

By the 1950s the Outlook Tower in Edinburgh was in a bad way and Arthur Geddes sold a large quantity of archival material to Dr Thomas Lyon of the Royal Technical College in Glasgow. Lyon gave the archive to the Department of Architecture. It passed on to the Department of Urban and Regional Planning and then to the Library of what was, by then, Strathclyde University. The Patrick Geddes Archive is vast (the Index alone consists of six weighty volumes). One marvels at the industry of the great man! And one can understand Mumford's despair at trying to bring order out of chaos.

Fortunately the impossible has been achieved and we can now access a mass of information with ease. The contents are very varied in their nature. There is a vast quantity of published material – books, maps, postcards, photographs etc – accessible elsewhere, but also much personal material, like correspondence, but often just scraps, barely legible but giving clues to the great man's thinking.

As for the remainder of the contents of the Tower they are now safely housed and carefully catalogued, and in normal times accessible to the public, in the National Library of Scotland and in the Research Collections of the University of Edinburgh.

There are items related to the Valley Section in the Strathclyde Archive, which we shall now consider, but there are difficulties. Some are mere scraps or scribbles. Some are barely legible. Most are undated. Nevertheless we can still trace the workings of a great mind. To identify them I have used Strathclyde's reference system.

'Exhibit 1'(T-GED22/4/31 and 22/4/32) is a bit of a puzzle. We have a provenance for the first – on the reverse side we have an endorsement in pencil – 'Cities Exhibition No 791'. They are two large sheets, on which are mounted water colour illustrations. The first is entitled 'Middle Valley Section' and has eighteen illustrations of rural occupations. The second is 'Lower Valley Section' and has nine illustrations or urban, rural and coastal occupations. Both sets were:

> Drawn & Etch'd by WH Pyne.
> Aquatint by L Hill.
> Published April 1, 1822 by R Ackerman, 101 Strand.

In the catalogue the illustrations have titles in French and English and some would fit nicely into a Valley Section diagram – Gleaners/Glaneurs, Ploughing/Labourage, Fishermen/Pecheurs Maritimes – but it all looks pretty tangential to the matter in hand.

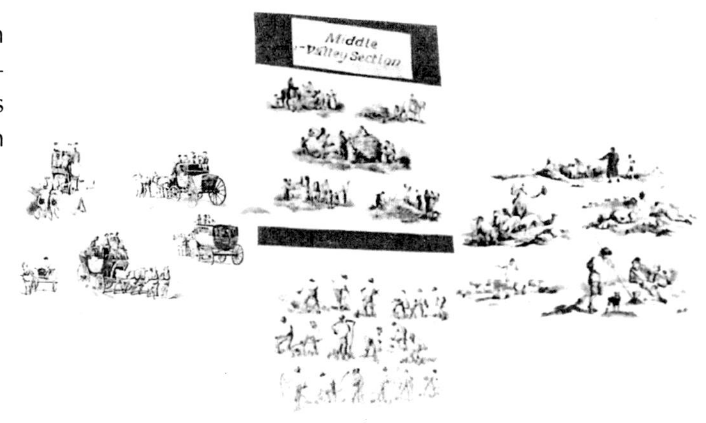

FIG 2
Figures from 1822 aquatint on
'Cities Exhibition No 791' Middle Valley Section
(University of Strathclyde)

Fig 2 is a selection of these attractive and authentic images and we can see in them harvesters and shepherds – but also coaches. It is difficult to see how these combine to illustrate the Valley Section.

T-GED22/2/11 and T-GED22/2/10 make an interesting untitled and undated pair. The former (Plate 4) is in colour (colour pencil to be pedantic, crayon on paper, cut and pasted on backing). Three peaks, like the Mountains of Mourne, 'sweep down to the sea'.

Seven conifers symbolise the Woodman and below them is the merest suggestion of deciduous trees. Again there are suggestions of a building in the 'farmland' and of a 'fisher toun'.

FIG 3
T-GED 22/2/10-pencil sketch (University of Strathclyde)

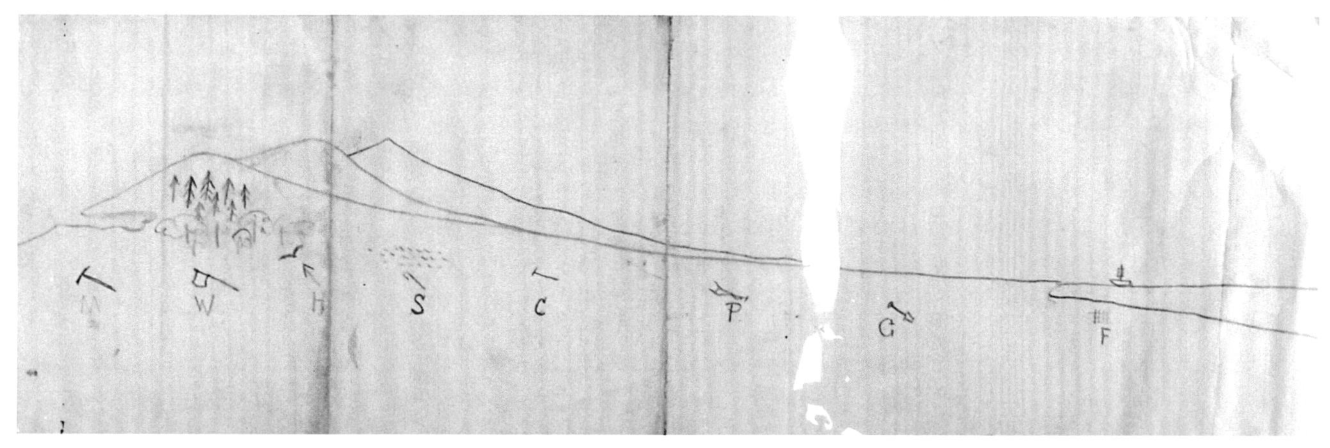

T-GED 22/2/10 is in worse condition, but is more interesting because it shows more. It is in pencil and on the back are barely perceptible notes and little sketches by Patrick Geddes. Looking at the front we see immediately the three hills (always on the left) sweeping down to the sea. We can see the same eight conifers, but the beech (?) trees are clearer. From left to right are shown eight occupations with their appropriate symbols. (The usual Valley Section has seven.)

Thus, we have M the Miner, with a pick above, W the Woodman with an axe under the trees, H the Hunter with an arrow aimed at a bird, S the Shepherd, with a crook and his flock above, C could be a Crofter or a Cattleman – with a rake? P is the Peasant, with a plough, G the Gardener with his spade and F is the Fisherman with his net and his boat floating on the waters above.

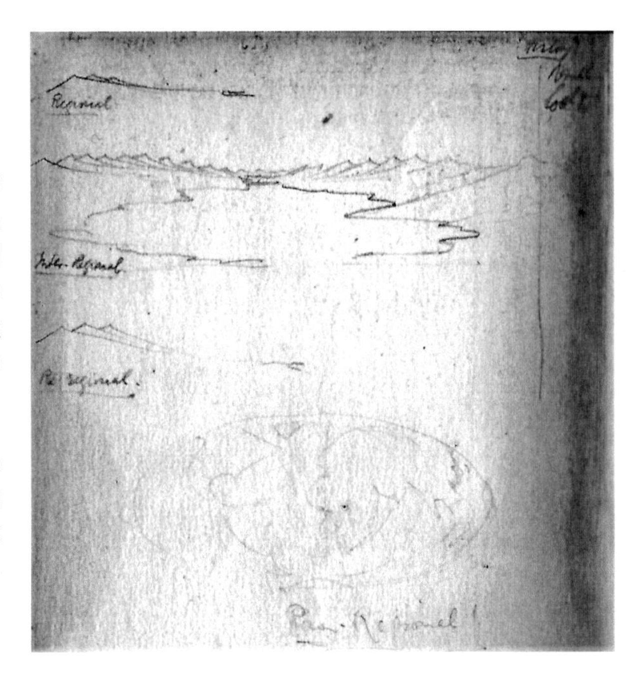

FIG 4
Geddes's sketchy Thoughts (University of Strathclyde)

In Geddes's own copy of GG Chisholm's *Handbook of Commercial Geography* of 1889 there are pencil annotations so faint as to be undecipherable. On one page there is the profile above, labeled 'Regional'. Below is another sketch – 'Inter-Regional'. Then the simple profile again, labelled 'P? Regional'. Nearby is a very faint sketch of the World on Mollweide's projection – labelled 'Pan-Regional!' (Geddes's emphases!)

When Kelman gave his lecture on *The Interpreter's House* in 1905 the stained glass panel of the Valley Section was in place in the Outlook Tower – and generally admired. T-GED 21/3/2 is a Post Card sent at 1030pm on 18th March 1908 to 'Professor Geddes, More House, Cheyne Walk, Chelsea'. (In which city/country? Postcode?)

FIG 5
Post Card sent by JB Mears, 1908
(University of Strathclyde)

According to Federico Ferretti - *'L'idée de la Valley Section est présentée pour la première fois par Geddes à une reunion de la Sociological Society de Londres en 1905'*. So this sketch, if it ever got into print, was not to be the first public appearance of the Valley Section. It shows three real hills with rain clouds overhead, a forest, a mine, and terraces - for the hunter? Then four sheep, a poor crofter's hut, a prosperous farmer's house set in arable fields, and buildings by the sea. No sign of Geddes's three birds, but an eagle or buzzard floats over the hills and an unidentifiable bird is over the fields.

Geddes, or his aides, was already tinkering with his concept.

Frank Mears was, in some respects, Geddes's right-hand man but was also a major figure in his own right. As Secretary of the Open Spaces Committee of the Edinburgh Social Union he was very active and sent Norah, Geddes's daughter, 'strange diagrams', and discoursed on art and symbolism. (They were married in 1916.) Mears did much of the graphic work for Geddes's schemes – exhibitions, photographic surveys, planning universities. As Captain Mears he served in the Royal Flying Corps under Major (Acting Lieutenant Colonel) Alasdair Geddes. Promoted to Major, Mears is credited with the invention of safety harness for parachutes. After 1918 he ceased to work in Geddes's shadow and went into practice on his own as a highly successful and respected architect and planner.

The sender of the postcard, JB Mears, Bartholomew or 'Barty', was the brother of Frank Mears. There was also a sister, Louisa. Barty trained as a doctor, was a bit of an antiquarian, and had a spell as manager of the Outlook Tower. Barty and Louisa may have joined the Geddes circle before Frank did. However, all three siblings became active at the Outlook Tower from around 1908, which was the year Frank returned to Edinburgh from London, where he had been working on designs for Khartoum Cathedral.

CHAPTER 7 SURVEY GRAPHIC
THE APOTHEOSIS OF THE VALLEY SECTION

PATRICK GEDDES PAID THREE VISITS to the United States, networking, picking up ideas and spreading the gospel. Lecturing was quite a money-spinner – on his second visit he picked up $1,200 ('not at all bad those days for a little-known botany professor'). On the 1923 visit Geddes was based in New York and was given a room in the New School of Social Research. He gave a series of lectures on Geddesian City and Regional Planning. He spoke to packed houses, although numbers tended to fall away. There was good sense and inspiration for those who attended – 'young architects and the usual vague sprinkling of females' – but who could discern it? Geddes's delivery was poor. The idea of a series of articles was mooted by the *Survey Graphic*, a social workers' weekly, and a stenotypist was brought in. The result was a 'terrible botch'.

Fortunately, a transcription of one of the lectures *(The valley section from hills to sea)* has survived which indicates two of the visual aids Geddes used in 1923.

They are:
Lámina 1. A perspective of a European Valley, and
Lámina 2. The Valley Section and its social types:
in their native habitat and in their parallel
urban manifestations.

FIG 6
A perspective of a European valley
(*Survey Graphic*)

Lámina 1 re-appears in the fourth of the Talks as a 'Bird's Eye View of Valley Region: Shewing some typical town sites and developments'. (Fig 8) The original (above) has explanatory labels, such as – Ford/City, Castle/Burgh, Ferry/Town, Seaside/Resort etc. – which were dropped for the second version. It is likely that Lamina 1 was Exhibit no13 in the 'Origins of Edinburgh' Section of the Cities and Town Planning Exhibition of 1910. Mabel Barker, Geddes's god-daughter, whom we will meet later, was responsible for the titles.

Strictly speaking, Lámina 1 is not a Valley Section, nor does it show 'social types'. It shows a strong family resemblance, not only to the 'Bird's Eye View' but to the 'Bird's Eye View of (the) Forth' by Frank Mears, which was also featured in the Origins of Edinburgh section of the 1910 Cities Exhibition. What these variations show is that, quite early, as early as 1910, Geddes's thinking had moved on from his simple little diagram.

Lámina 2 has a chapter to itself - later.

As for the lecture the introductory section bears a strong resemblance to the beginning of the fourth talk and then goes on to a detailed examination of each of the social types. Geddes concludes:

> In short, our geographic and historic surveys are increasingly yielding us a philosophy, an ethics and a policy of social life, in which all that is best in the various divergent schools of thought and action may increasingly work together.

Geddes's core material was too good to lose, so he revised and rewrote until, in 1925, six articles appeared monthly in *Survey Graphic* as 'Talks from My/The Outlook Tower'. (The series started as 'My' and changed quickly to 'The' Outlook Tower.)

Geddes's drift can be gathered from the titles of the articles:
A Schoolboy's Bag and A City's Pageant
Cities and the Soils They Grow From
The Valley Plan of Civilisation
The Valley in the Town
Our City of Thought
The Education of Two Boys

The essays were in effect the outpouring of all Geddes's ideas on society in the past and in the present and a rambling summing up of his life and achievements. As such they have been an accessible source – particularly *The Education of Two Boys* - for most of his followers. At the time his audiences must have thought these were his farewell thoughts – he was almost seventy – but they were wrong, he still had a couple of cards up his sleeve!

Our concern is his Valley Section and, as can be seen, he devoted two lectures to it. We do not know precisely what visual aids he used in the original public lectures, but he had also had illustrations made for his lecture tour incorporating modifications and elaborations, and these made their way into the *Survey Graphic* collection.

A Schoolboy's Bag and *A City's Pageant* were favourite topics for Geddes (and, in fact, a Schoolboy's Bag was one of the decorative/didactic features incorporated into Burns' Land, a Geddes student hostel in Edinburgh's Lawnmarket). It is an essay in concentric education. As the items are emptied out Geddes links them with the whole of creation. Educational historical pageants and masques were another way of engaging local people with their past. For example, when he wanted to show the sterling qualities of our ancestors, he did not show Wallace, Bruce and Bannockburn. For him the one incident in military history he repeatedly saw as exemplifying loyalty and internationalism was the service of the Scots Guard of Joan of Arc in her struggle to free France from English rule, when her Scots bodyguard entered Orleans and probably Rheims for the coronation of the French king.

Geddes on the move was as active as the modern businessman on his laptop, and in *Cities and the Soils They Grow From* he writes:

> As I revise this manuscript, our ship is running along the east coast of Sicily, between Messina and Catania – a noble landscape, with Etna towering behind the minor mountain-masses which rise from near the shore.

Geddes knew his Mediterranean. He had studied in Naples and in 1897 had surveyed and reported on Cyprus, had formed the Eastern and Colonial Association Ltd, bought two farms and proceeded to demonstrate how Cyprus could be changed. He also had had a season designing gardens on the Riviera for *les hivernants* – refugees from the northern winters.

The lecture proceeds to take a description of the coastal detritus and malarial breeding grounds of the Mediterranean and link them into a network of deforestation, over-grazing, soil erosion and urban decline.

'The Third of the Talks from the Outlook Tower' was published in June 1925 and was headed up by a very simple, basic, and by now very familiar Valley Section. Three hills sweep down to the sea and underneath we see the titles of six occupations and the symbols for eight. Geddes has already started refining his initial model.

We see the Miner digging in the mountains, the Woodman with the conifers, the Hunter in the deciduous woodland and the Shepherd in the rough pasture. But he has three types of Peasant, a poor one with a mattock and a miserable crop, an arable farmer with a plough and a strong crop, perhaps of wheat, and a market gardener intensively working the land with a spade. The two Fishermen are well placed for selling their catch in the town on the horizon.

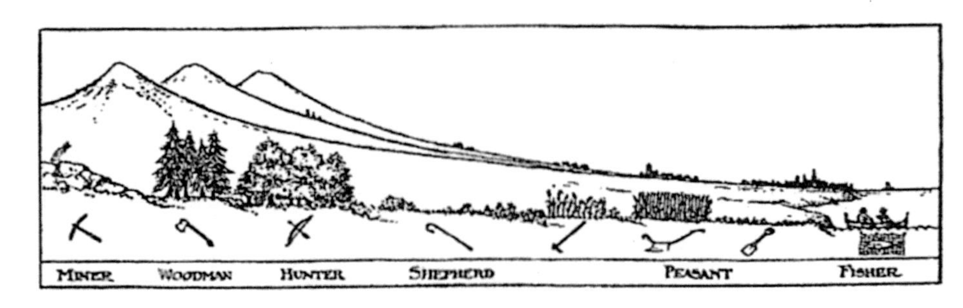

FIG 7
The basic Valley Section (*Survey Graphic*)

In *The Valley Plan of Civilisation* Geddes tells us that:

> In the social sciences...the essential is to have as clear an understanding as we may of normal life-processes before we come to pathological interruptions. So before coming (or going) to War let us learn more of the ways of Peace.

Geddes starts by taking corn for the West and rice for the East and teasing out what he calls the civilisation-values of our staple crops. He then invokes the Valley Section, relief, the climate, the soil, history, to establish the basic factors which influence man's use of the land. (Influence, not determine.) In turn he takes each of the nature-occupations and describes and analyses the factors that contributed to their establishment in particular segments of the Section. But he is not deterministic and makes it quite clear, at the end of the lecture, that although these fundamental occupations are always with us, they are also continually in a state of flux. In fact, the next lecture examines how the Valley Section is affected by urbanisation, in particular.

In his lecturing style, Geddes displays his adherence to the French romantic school of geography. Not one Germanic statistic in sight. Instead we have a wide-ranging and eclectic range of generalisations and a desperate groping for connections. There is no question that his message is lively, interesting and occasionally mind-blowing, but he tries too hard, on occasion.

What can one make of the following, in the context of the Shepherd and the Hunter?

> Thus of old Abraham and Lot separated when their young men quarrelled; and again, in recent times Kruger delayed while Chamberlain and Rhodes pressed on for the Boer War.

Under the Fisher we find:

> Indeed it is thus by no mere accident, but also from deep-rooted tradition that my old and honoured friend, the veteran president of the International Council of Women should bear the title of Marchioness of (the old fishing port of) Aberdeen.

Really? Ishbel Marjoribanks (1857-1939) was born in Belgravia, where the first Lord Tweedmouth, a brewing banker, had a 'palace'. He also had 19,186 acres at Guisachan in Invernessshire. She married Lord Aberdeen in St George's, Hanover Square, London. Active in Liberal politics the Gordons' life was divided between the House of Lords and Haddo House – twenty miles from Aberdeen. Ishbel had two spells as Vicereine in Ireland (where she brought Patrick Geddes and his exhibitions over for two extended visits and where Norah Geddes stayed on to work on garden projects). Her relationship with Geddes was more than professional. There is a warmth in their correspondence when she addresses him as 'Uncle Pat'.

She supported her husband when he was an enlightened Governor-General of Canada and from 1893 she was all over the place as President of the International Council of Women. In short, for most of her life, Aberdeen was merely somewhere the Marchioness passed through on the way to somewhere else. However, seven years after Geddes's death, in reduced circumstances, she died of a heart attack on 18 April 1939 at Gordon House in Aberdeen.

The lengthy narrative and analysis of the seven rural types had already been used, almost word for word, by Geddes in several contexts and would be used again, as we shall see.

For the fourth lecture, *The Valley in the Town*, clearly the emphasis has shifted. The Valley Section is still valid but has been elaborated to take account of change, particularly industrialization and the growth of towns. This is best illustrated by a consideration of the four illustrations which accompany the lecture.

FIG 8
Bird's Eye View of Valley Region (*Survey Graphic*)

A 'Bird's Eye View of Valley Region: Shewing some typical town sites and developments' is clearly not a Valley Section, although we are looking upstream, but can be recognized as a development from it. Lamina 1 (Fig 6) is the same basic drawing but has labels in appropriate places – seaside resort/ford city/ fishing village/country town etc – which the Bird's Eye View has not.

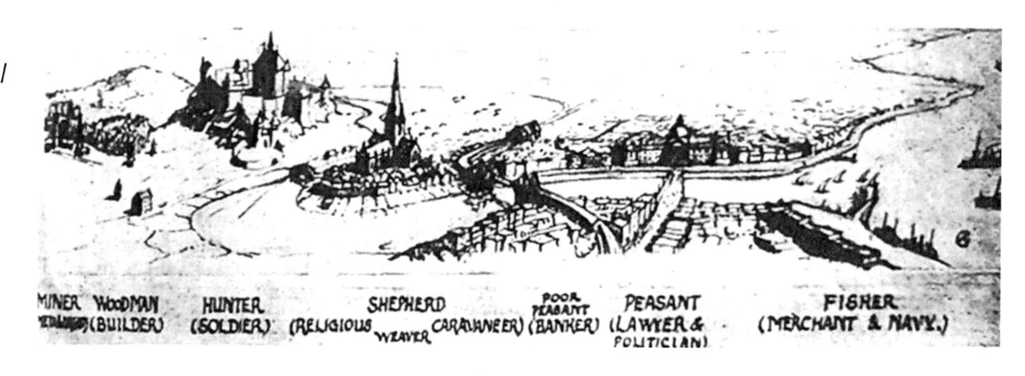

FIG 9
Aux origins de l'aménagement regional
English version (*Survey Graphic*)

The next illustration (Fig 9) comes from a 1925 paper – *Aux origins de l'aménagement regional: le schema de la Valley Section de P. Geddes* – but was also in the fourth published talk. Although this has the usual hill profile and there is a river, it is really a town study - and has little sketches which fill out the commentary below.

The captions illustrate change. Seven 'occupations' become nine and the text of the lecture elaborates these changes in great detail. For example, the military relate to the castle, or a strongpoint, which benefits, for strategic reasons, irregularities in the terrain. Fishermen and seamen are found by the estuary which has benefited from construction of river embankments and harbour works.

What I call 'the double section' (Plate 5) was first designed for Geddes's Cities Exhibition of 1910 and was used as an illustration to the fourth lecture. Its official title is 'The Valley Section and its social types: in their native habitat and in their parallel urban manifestations'.

Let Geddes himself explain it to us. Below the upper picture comes:

> From the hills to the sea the valley section discloses miners, gold-washing in the stream; woodmen destroying the last forests; shepherds; patriarchal hunters and their clans; peasants reaping; rich farmers; fishing village and boats.

And under the lower picture:

> Starting from the left, the iron works and the shops of the ironmonger and goldsmith (miner) appear in the city street; then (woodman) timber and fuel yard, furniture shop, paper warehouse. Next the big ('caravaneers') store with furs from hunter and woolen (sic) goods from shepherd. Also there is his small church or chapel with which there should be a school. The Hiring Fair, the original of the Labour Exchange, indicates the older appearance of the poor peasant, and in later times we see his bank and insurance company. The farmer occupies the street as miller, baker, brewer, butcher, greengrocer, innkeeper etc. For the fisher is the fish shop and for the sailor and merchant venture the warehouse and shipbuilding yard and the great seaports.

The text of the lecture is pure Geddes, a mélange of good logical thinking and practical experience, unsubstantiated assertion and doubtful sociology. Section V is the last part of the lecture and in it he rambles around recapitulating all that has gone before. What are we to make of this incident from Scottish history?

> Hence too King Robert the Bruce's initial stroke at Bannockburn – warding the English champion's charge, and cleaving him as he thundered past – avows his woodman-fisher origin, by his expression – not of pride or joy over this omen of his victory as both armies saw it, but of disgust at himself on noticing his notched blade – 'I have broken my good battle-axe!' – a feeling that every lumberman knows.

FIG 10
The Eastern Watershed of Scotland (*Survey Graphic*)

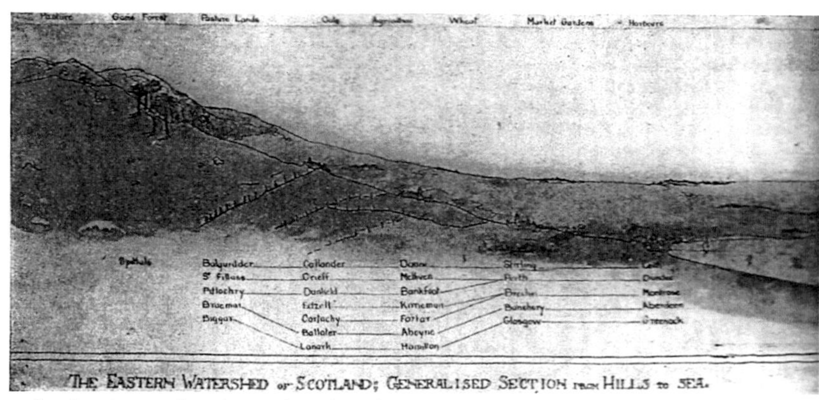

Introducing this section we have 'The Eastern Watershed of Scotland: Generalised Section from Hills to Sea' (Fig 10). Superficially, this is a Valley Section. There are the hills on the left and the sea on the right. Along the top we have eight environments, not occupations – Pasture, Game Forest, Pasture Lands, Oats, Agriculture, Wheat, Market Gardens and Harbours. Then we have some of the usual sketched landscape features – trees, sheep, dwellings, etc. Next comes an array of place-names, arranged in columns below the environments. 'Spittal', like the Spittal of Glenshee, was a hospice or mountain refuge; the others are settlements big and small, which Geddes has arranged in groups. Each group could be made to fit into a Valley Section. This is really quite ingenious and when we examine each place we can see what he means. But there are problems.

The figure runs from high on the left to low on the right, from west to east. But the bottom section runs from Biggar, 28 miles south of Edinburgh, to Greenock, very definitely in the west. In reality the bottom section is really a mirror image of the section above.

More seriously, after much scrutiny, I still cannot find out where and how it fits into the last part of Geddes's lecture. It is as if Geddes had said 'Here's an interesting thing, let's find a home for it somewhere.'

> Having now disposed of the Valley Section as a topic, to his own satisfaction, Geddes now moves on to *Our City of Thought and The Education of Two Boys*.

CHAPTER 8 CARTOONS BY HENDRIK WILLEM VAN LOON

A CARTOON IS NOT NECESSARILY FUNNY. The writer's dictionary has, as its fourth definition of 'cartoon': 'a full-size preparatory sketch...from which the final work is traced or copied'. In other words, a cartoon is not a finished work, but a work in progress; and in calling his illustrations cartoons Van Loon acknowledged that he was a contributor to a process of dialogue and persuasion. Van Loon was born in Rotterdam in 1882, looked at the world through Dutch/European eyes, but spent most of his life in the United States. In his youth he found Holland narrow and censorious, prejudiced and concerned with the niceties of religion rather than its central message.

At twenty he went to the United States out of sheer curiosity, graduating brilliantly in history at Cornell in 1905. After a year at Harvard he became a journalist, covering the Russian Revolution of 1906-7. Then came Munich and a PhD in 1911, university teaching in America, and travel all over Europe in 1914-19 as a war reporter, the Netherlands being neutral. During the Second World War he broadcast to occupied Holland from Boston and worked on his varied humanitarian concerns – aiding refugees from Nazi persecution and war-relief fund-raising. He died in Old Greenwich, Connecticut in 1944. He did not live to see the liberation of Veere in Walcheren in November 1944.

His main achievement was a steady flow, starting with *The Story of Mankind* in 1921, of major popular texts – ostensibly aimed at children. They gave an approach to history that enlightened parents and teachers had been needing. *Van Loon's Lives* was published in 1943. On the face of it this is a collection of essays on important historical figures, sugared by being set, imaginatively, in the town of Veere, but is really a justification and a plea for liberal democracy in a world dominated by fascism.

The book was dedicated to Juliana (later Queen of the Netherlands) and:

> ...those valiant men of our beloved Zeeland who died while trying to preserve and maintain that most cherished of their possessions their LIBERTY. (Van Loon's capitals).

Van Loon illustrated his own books. The Rembrandthuis in the Jodenbreestraat of Amsterdam has no great biblical scenes, no great portraits in oils, no *Night Watch*, just intimate pencil drawings and little sketches using pen and ink. Often no more than a few ink scrawls on a piece of white paper, by some miracle the white seems to have three dimensions and the result is a miracle of expression. Van Loon quite clearly took Rembrandt as his model.

For Geddes, Van Loon was much more than a 'name' to be brought in in 1923. Twice he is mentioned in the *Talks from the Outlook Tower*. Geddes says:

> We are thus brought face to face with the enormous and laborious studies of histories, literatures, origins to which Van Loon or Wells give excellent introductory primers.

And again:

> Wells' *Outline (The Outline of History*, 1920) seems to be having a widely educative result, on both sides of the Atlantic, in abating too nationalistic sympathies; as also Van Loon's and, of course, more specialised works.

FIG 11
Five Van Loon illustrations (*Survey Graphic*)

The elementary division of human labor

When the trees are gone . . .

The city could not but decline

In this case our conversion was to the new and frightful toys of experimental chemistry

. . . a long range of noble precipice

Fig 11 has five of Van Loon's illustrations. The top row very effectively relate to Geddes's constant theme of the degradation of the environment by deforestation and soil erosion. The lower two relate to *The Education of Two Boys* and authenticity. Geddes's father set the lad up in a workshop/laboratory. Did Van Loon and Geddes sit down together and list what was to go into the sketch? Or did Van Loon just go ahead and draw what he thought was appropriate? And did Geddes approve of the result?

'A long range of noble precipice'? Kinnoull Hill is heavily wooded, with a mock-Rhenish tower on top. The south face is certainly cliffed, but below it is the Tay and the Firth of Tay – with mirrored hills only a couple of miles away. No vast amount of water and a lone steamer. Van Loon must have concocted this scene in his New York studio. Geddes could not have approved its publication. Is this an indication of his hectic lifestyle?

FIG 12
Six cartouches by Van Loon (*Survey Graphic*)

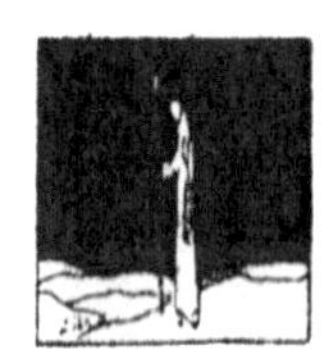

When it comes to Fig 12, we are on safer, non-controversial ground. These are simply little cartouches, each at the head of a new section of the text.

From left to right, they are associated with:

The Miner (also Woodman and Hunter) – The Shepherd (the only night-time scene) – The Fisher – War-origins – Cereals – Assyrians.

PLATE 1
Murdo Macdonald
deconstructs the Valley Section.

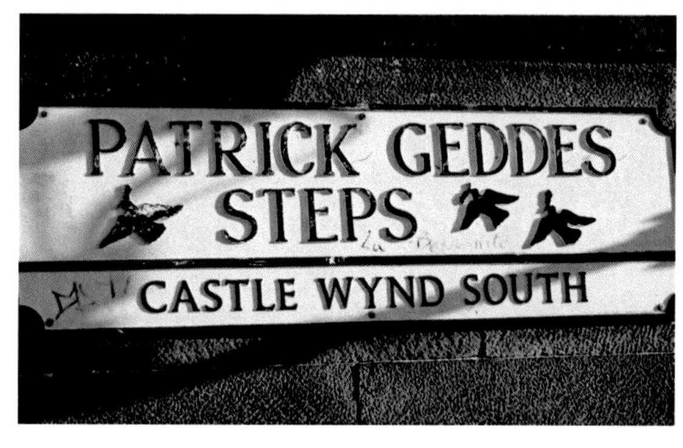

PLATE 2
Patrick Geddes Steps,
Castle Wynd South
(N.B. 3 birds with oak leaves).

PLATE 3
Section through the Tay Valley
at Woody Island.
(Perth Museum and Art Gallery).

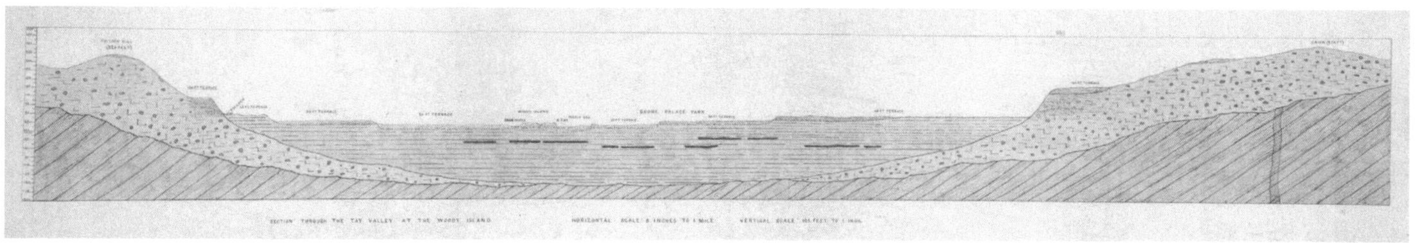

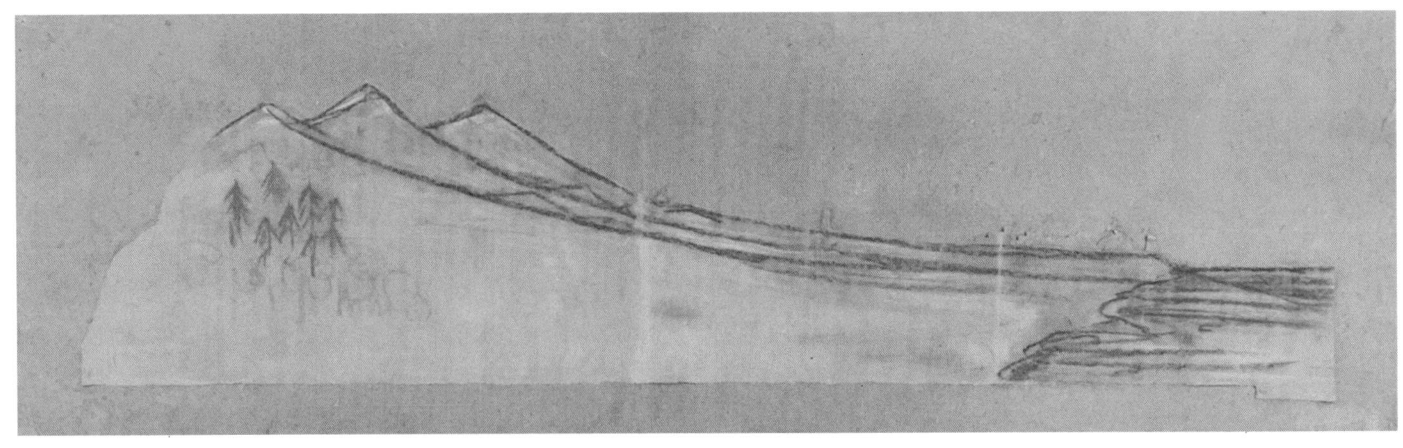

PLATE 4 T-GED22/2/11 – undated sketch (University of Strathclyde).

PLATE 5 The Valley Section and its social types. The ‘double section’ (University of Edinburgh).

PLATE 6 Die isolierte Stadt and Modification.

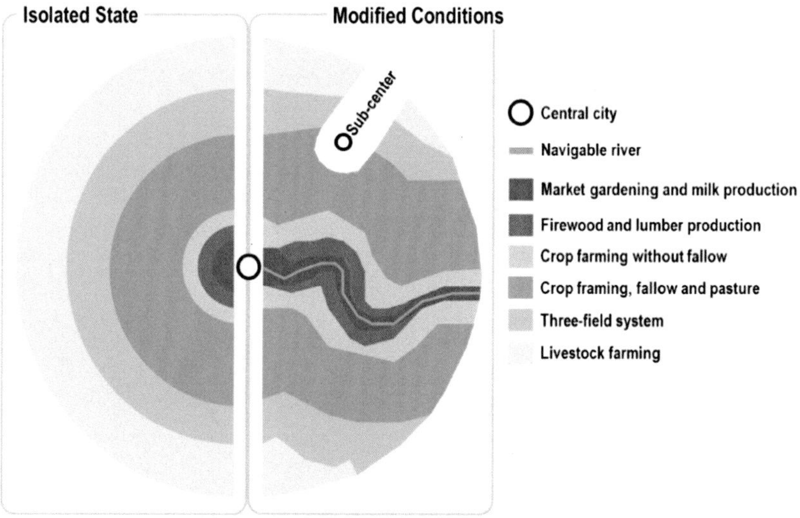

Isolated State

Modified Conditions

Sub-center

Central city

Navigable river

Market gardening and milk production

Firewood and lumber production

Crop farming without fallow

Crop framing, fallow and pasture

Three-field system

Livestock farming

PLATE 7 Leadership or Co-ordination (The Coming Polity).

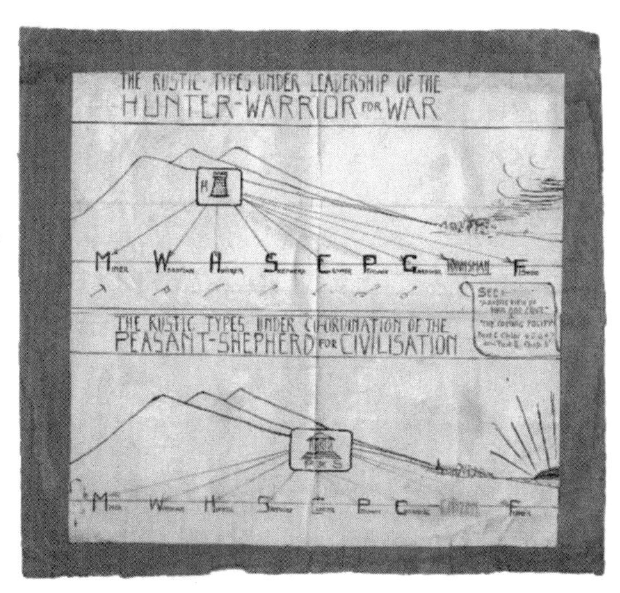

PLATE 8 A 21st century Valley Section (Dirty Teaching).

PLATE 9 Conference Programme
Yamaguchi Valley Section
Reflections 2018.

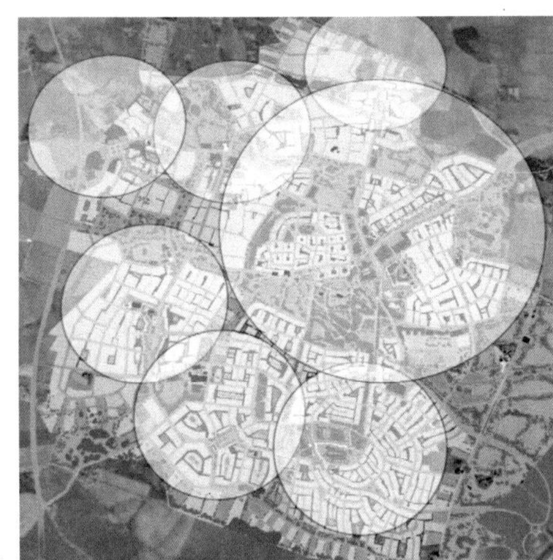

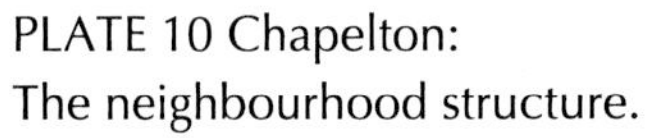

PLATE 10 Chapelton:
The neighbourhood structure.

CHAPTER 9 THE SIXPENNY VALLEY SECTION

IN THE OLDEN DAYS, TEACHERS, exhausted after the summer holidays, took refuge by setting the whole class, as an essay subject, 'The Story of a Sixpence'. A sixpence was the smallest of the silver coins, one fortieth of a pound sterling, and, as such, had a certain resonance. There was a popular song – 'I've got sixpence...There is tuppence to spend, and tuppence to lend and tuppence to send home to my wife'. 'Half a Sixpence' was a stage show and film built around the use of the coin as a love token. (In Scotland it was a 'silver bawbee' – two pence).

Plate 5 - 'The Valley Section and its social types: in their native habitat and in their parallel urban manifestations' has already been mentioned. As Lámina 2 it was an illustration for the Valley in the Town lecture in Survey Graphic. The accompanying text was given earlier. I know it as the 'double section'. It is now, nicely framed, in Edinburgh University Library. The question is – why? And how did it get there?

The double section was created in 1910, a year of exhibitions. While Geddes probably created the design the meticulous draughtmanship was probably the work of Frank Mears. Lord Aberdeen had two spells as Viceroy in Ireland. John Sinclair, a radical Liberal and Home Ruler, was his Aide-de-camp and secretary in his first spell and followed him to Canada before returning to a political career in Britain. In 1904 he married the Aberdeens' elder daughter and served as Secretary of State for Scotland from 1905 until 1912. As Secretary of State, in 1910 he officially opened the Geddes civic exhibition in Edinburgh. The Aberdeens were present and were so impressed by Geddes that he was invited to visit Ireland to support Lady Aberdeen in her development programme to head up civic exhibitions, medical inspections, civic surveys and the like – twice. Geddes was particularly welcome because he could penetrate into areas and interview people the representatives of Imperial power could not. After the first visit Norah, Geddes's daughter, stayed on in Dublin to head up a programme of community gardens and playgrounds.

After 1911, in Edinburgh, Dublin and Belfast the double section was part of Geddes's Cities and Town Planning Exhibition which won international acclaim. At the great 'Exposition Internationale' in Ghent in 1913 the international jury awarded the *Grand Prix* to Patrick Geddes. As he said: 'The Germans may have the frames and labels, but we have the ideas'.

Meanwhile, in 1909, Sinclair was created Lord Pentland and in 1912 was appointed Governor-General of the Madras (Chennai) Presidency in India.

At the outbreak of the First World War, the exhibition – with the double section – was in the Linen Hall in Dublin. One evening, after dinner, Geddes was vigorously declaiming:

> I urge not only that the collections in the Linen Hall be preserved permanently, but that the replanning scheme which wins the Viceroy's prize can be carried out courageously and beginning this very summer.

Later that same evening, Lord Aberdeen, having received a message, broke the news that war had started. It was 4 August 1914. Next Sunday the Exhibition was dismantled for the Linen Hall to become, first a barracks, then a training centre for nurses, then a hospital. On the fifth Geddes typed a remarkably perceptive letter to Alasdair. He saw the war as: '...the fruition of past ideas and ideals and their applications: and it must now run its course; as such things do'. He foresaw the collapse of the Austrian, German, Ottoman and Russian Empires, revolution in Russia, mutiny and civil unrest in Germany and Austria, 'perhaps even in minor cities like this'. He said 'the exasperated and angry people of every capital' would seek 'to do away with their incompetent rulers'.

In April 1914, Geddes had contracted to take his Cities Exhibition – with the double section – to the Madras Presidency in India, with himself as planning consultant and Alasdair as assistant. On 6 September Geddes and Alasdair sailed for India on the P & O liner *Nore*. The Cities Exhibition was packed up and despatched separately on the *Clan Grant*.

'The Cruise of the Emden' or 'The German Sixpence' would qualify as stories for the Boys' Own Paper. Germany had an overseas empire in Africa and around the Pacific. On 14th August the German East Asia Squadron set off for Germany to outgun, outsail and totally defeat a British squadron at Cape Coronel off the coast of Chile, only to be destroyed in turn at the Falklands. Meanwhile Captain von Müller took the light cruiser *Emden* into the Indian Ocean, followed by four colliers (ships, not miners) so that he need not put into any port for refuelling.

With wireless telegraphy still in its infancy there now began a gigantic game of Blind Man's Buff in the vast spaces of the Indian Ocean. There would suddenly appear over the horizon this four-funnelled cruiser (von Müller had added a dummy funnel to make his ship look like HMS *Yarmouth*), which would fire a shot across your bows, hoist the German ensign and signal: 'Stop at once. Do not wireless'. The crew and passengers would be taken off and the ship sunk. Neutral vessels would be released and given some of von Müller's 'guests' to take back to port.

On the night of 22 September *Emden* shelled the oil tanks of Madras and on 28 October she sailed at full speed into Penang harbour, sinking the *Zhemchug* (a Russian cruiser which had been sent to find and destroy *Emden*). The captain of the Zhemchug had been ashore seeing his mistress during the attack; he was subsequently demoted and imprisoned. A party of 60 Chinese prostitutes were aboard at the time of the attack; their fate is unknown. *Emden* was pursued by a French destroyer, which was itself destroyed.

Panic reigned around the Indian Ocean. Insurance rates shot up. It was a source of great embarrassment to the British and other Allies that a single German cruiser could effectively shut down the entire Indian Ocean. Looking for *Emden*, the tiny needle in a vast haystack, were four vessels of the Imperial Japanese Navy, three French, two Russian and the British *Hampshire* and *Yarmouth*, HMS *Weymouth*, RMS *Empress of Russia* and SS *Empress of Australia*.

Von Müller took his ship to the Cocos Islands to destroy the Eastern Telegraph Company wireless station. Fifty armed seamen went ashore and knocked down the radio tower (showing consideration for the tennis court!) – but not before a general call went out.

A mere 55 miles away the Australian light cruiser HMAS *Sydney* was escorting a convoy. Within three hours she had engaged the *Emden*, who had had to leave the boarding party ashore. Outgunned as *Emden* was, the battle lasted three hours till she was beached to avoid sinking. After further bombardment she surrendered having lost 131 dead and 65 wounded. The shore party and the German prisoners went on to further adventures – but that's another story!

Over 20 British ships were sunk by the *Emden*. On 16 October three British ships were sunk, including *Clan Grant* of 3,948 tons, with Geddes's precious exhibition aboard. The Cities Exhibition was a main source of Geddes's income, with the Double

Section as his main visual aid. A schoolboy's story might have ended here, or a hero like Wilson of *The Wizard* might have invented a diving machine and plumbed the depths of the Indian Ocean, but was this the end of the Double Section?

A lesser man than Geddes would have crumbled but, as he wrote to Lord and Lady Aberdeen:

> The best I can say is that having raged so much over Louvain and other cities, I have not lost sleep over the loss of even the best of plans and pictures of them. Still it is hard...to lose years of work!

However, something like a miracle happened. Back in Britain an Emergency Committee of friends – including Arthur Geddes – headed up by HV Lanchester, a planner, rallied round to stitch together a replacement Cities and Town Planning Exhibition. This was shipped out to India in time to open in Madras on 17 January 1915, after Alasdair had worked frantically on its display. Consider the achievement. 93 days after Mark I Double Section was sent to the bottom of the Indian Ocean, Mark II was on display in Madras – without electronic communication and processing and air transport!

Geddes had three extended spells in India between 1915 and 1923, in 1919 (having retired from Dundee) becoming Professor of Civics and Sociology at Bombay (Mumbai). He is said to have produced over fifty surveys or reports for the improvement of Indian cities. A Geddes project was no simple matter, definitely not a stroll round the town with a clipboard and pencil. The survey was an event and a process. There was the active part of the survey – Sympathy of his planning model. Collecting data, measuring and counting, interviewing local people. But there was also an event, the staging of a season of activity, the viewing of the Cities Exhibition, lectures, discussions, walkabouts. As for the Cities Exhibition, it was the same for each city, except that, for the last room, the local survey and plan were displayed (Synthesis), part of the process of creating Synergy – the carrying out of the plan with the participation of the local people.

As we have seen Geddes visited the USA in 1923, lecturing. One of his visual aids must have been a lantern slide of the Double Section. In July 1925 the *Survey Graphic* illustrated the *Valley in the Town* lecture with the Double Section. Where did the New York printer obtain the master copy from?

In 1924 Geddes left Bombay for health reasons, moving to Montpellier where he established his Collège des Écossais. The area was a perfect example of the Valley Section, with rural types from the Cevennes down to the Mediterranean. (See Fig 20). In 2014, Patrice Bouche of the University of Lille wrote:

> The site for the college was chosen because it matched quite perfectly some core conceptions in Geddes's thought... the Lower Languedoc...was a life-size model of Geddes's concept of the 'Valley Section'.

The Cities Exhibition was brought from India to form an important educational aid in Geddes's enterprise. He now had two financial windfalls. He was awarded a Civil List pension of £80 (= £8,000 per annum today). The Royal Navy was the ultimate guarantor for all British ships and cargoes. The Navy had failed to protect Geddes's exhibition so, in 1926, the government War Damage Agency awarded him £2,054 as compensation for its loss. Which sum was immediately swallowed up in his new venture.

Geddes died in 1932 but the Scots College kept on going until the Second World War. However, in 1934, probably at Arthur Geddes's insistence, the Cities Exhibition was transferred to Edinburgh, to 60 George Square, to the Edinburgh University Department of Urban Design and Planning. Here the Double Section was displayed in a corridor and used as a teaching aid with the students.

On 5 March 1985 the Department – now the Department of Urban Design and Regional Planning – was moved to Chambers Street. In the Matthew Architecture Gallery was held, from 16 September to 22 October 2004, an exhibition – 'Patrick Geddes: The Regeneration of Edinburgh'. Material from the Cities Exhibition and from the Outlook Tower – including our section – was brought together. On 2 October (150th anniversary of Geddes's birthday) was held an all-day symposium at which was launched *Think Global, Act Local: The Life and Legacy of Patrick Geddes*.

Our Sixpenny Section, beautifully framed and conserved, is now in the Special Collections of Edinburgh University Library, back in George Square. After long periods as a stimulus and visual aid it is not on display, but can be made accessible to interested parties on application.

Our Story of a Sixpence has been full of drama and uncertainty but certainly has a happy ending. It is a sobering thought to consider how many people, in several parts of the world, and in over a century, may have been affected by it and its message.

CHAPTER 10 FAITH AND THE RIVER

A RIVER IS A GOOD FRIEND BUT A BAD MASTER is a commonplace that still has some credibility. A river can bring us clean water for our daily needs and supply us with food. In a country like Egypt the Nile shows up as a ribbon of green running through the pitiless desert. Any one river can be both friend and master. Transport up and down the river is cheap and easy, but the river is often a barrier to be crossed at great expense or cause expensive diversion or costly delay.

In the case of the Nile, for most of its life it has provided water to grow crops along its course. But every year, a thousand miles away in the Abyssinian Highlands, torrential monsoon rains turn the Blue Nile into a mighty torrent which floods the farmed lands of the lower Nile, destroying everything. But when the waters recede the cropland has been refreshed and reinvigorated by the silt brought down from the highlands. Of recent years, of course, the Nile has been 'tamed' by the construction of great dams which regulate the flow of water and generate electricity. But in the process the enriching silt is deposited in the reservoirs behind the dams so that the *fellahin* have to have recourse to imported fertilizer for their cotton and other crops.

The need to keep a calendar in order to predict the arrival of the annual flood and the need to keep measurements and records of land allocation for reconstruction after the floods – and for a central organization to manage these – are factors which have led to Ancient Egypt being seen as a 'Cradle of Civilisation' like the Tigris and Euphrates in the Middle East.

For many, the river is not a silvery carpet sliding quietly down towards the sea. For many it is a raging monster that appears from nowhere and carries all before it, sweeping away trees and humble dwellings, cattle, and often people, and covering good land with gravel. Mysterious and hostile, is it surprising that our ancestors thought that something bigger than mere humanity should be at work among them? And should personalise these forces, try to placate them and conciliate them when all is going well?

Fig 19 shows some of the big players who were gods to the Greeks and Romans but our ancestors had a plethora of local gods of the river or of the well. Faint traces of these beliefs survive in the few surviving 'clootie wells' in Scotland or with those who cast coins into the Trevi fountain of Rome – celebrated in song and film.

As we have seen, Geddes was born into and brought up in the Free Church of Scotland. Although he lost his faith he still saw the Free Church as an influential force in society and had many supporters from it. He kept an open mind and even invited the suspicion of the local establishment by giving attention to purveyors of strange belief systems, like Annie Besant and Helena Blavatsky of the Theosophical Society and the leader of the Bahaí'í faith. Although he was not to set foot on India until 1914, at age 62, Geddes began associations with Indians in America and Britain – notably Nobel prizewinner Rabindranath Tagore (1861-1941) - from the 1890s.

Sister Nivedita (née Margaret Elizabeth Noble of Dungannon, County Tyrone) worshipped Geddes from afar until they met at the International Association of Religions in New York in 1900. This morphed into the Congress of the History of Religions, which ran in parallel with the great Paris Exhibition of 1906. Sister Nivedita was invited to be Geddes's secretary for these events. But it did not work. She felt she was only given menial tasks to perform and could not contribute ideas or experience of her own.

On arrival in India in 1914, Geddes found Annie Besant solidly entrenched there. Despite her incessant foreign travel, the international headquarters of the Theosophical Society were now in south India and in 1917 she was elected as President of the 32nd Indian National Congress – not bad for a white woman! She drew Geddes into association with Gandhi, so that each of the three was marching in roughly the same direction.

For Besant 'the river' represented a key symbol for regeneration, which lies at the heart of Hinduism. The river echoes the flowing, meandering thread of life – or life echoes the flowing meandering thread of the river.

> Besant recognised the essential spirit of India and its rivers as enabling the renewal of the spirit and the soul – a recurring theme in the poetry of Tagore.

Kenneth Munro, Chairman of the Sir Patrick Geddes Memorial Trust has consistently examined rivers and their effects on humanity. He explored the Murray River in Australia in 2000. In 2004 (150th anniversary of birth of Geddes) he headed up a project *The Language of Rivers and Leaves* (Patrick Geddes – 'By Leaves We Live'), involving Bengali and Ballater communities. In Kolkata, students of the School of Art and Craft, with a local boat yard, constructed, by traditional methods, an 18-foot Bengal riverboat. She was named *Sonar Tari* (Golden Boat). This galvanised an exercise in global social awareness.

There was a pageant, Geddes-style, radio commentary, singing and dancing. The boat was filled with models, flutes, masks and paintings by children and students, crated and shipped to Scotland, via Colombo.

From 1999, Ballater and its primary school had been busy with tree planting, a timber trail with metal markers, a 'concept bus shelter', a bronze plaque on the site of the Geddes birth-house and an exhibition. Now, in Aberdeenshire, a programme of creative exchange brought about Indian music, dance and model boat making workshops with a broad cross-section of communities. The *Sonar Tari's* journey was completed 'by juggernaut and tractor'. The boat, which started on the Hooghly, was not taken to the Dee (too dangerous – 'The Dee, the Dee, each year claims three'), but to Loch Kinord, where it was launched and baptised. Pulak Goshe, Director of the Kolkata school and Kenneth Munro were able to navigate Indian wood on Scottish water.

The boat was displayed for a month in Alford Transport Museum. An exhibition of the project (sans boat) was displayed in the Lobby Garden gallery of the Scottish Gallery and was well attended. Robin Harper, the first Green member to be elected to a British Parliament, presented to the Parliament the formal registration of the ongoing significance of the philosophy promoted by Sir Patrick Geddes. Sonar Tari then went into dry dock at the Scottish Maritime Museum at Irvine and is now at the new ECO Centre on Kinghorn Loch in Fife.

This project, in turn, led to a collaboration with Bashabi Fraser (another PGMT Trustee) in her publication *From the Ganga to the Tay*.

The following year Munro was again in Kolkata, this time setting up *The Song of the Rickshaw*. He witnessed 'the extraordinary festival venerating the Hindu goddess Durga'. Thousands of temporary shrines, painted, festooned with fabric, in replicas of historic temples, are created.

> On specific days, in seemingly endless progression, thousands of these artworks, some three metres in height, are moved by truck, ox-cart and rickshaw to the banks of the Hooghly, where teams of young men struggle and strain to carry the sculptures to the water's edge for immersion and dispersal.

The event is cleansing, liberating and very much in the spirit of Hindu philosophy: celebrating birth, death and renewal by uniting the Ho river as part of the mighty Ganges (Ganga). On an economic level, this annual creative ritual also ensures that artists and crafts people have a cyclical series of commissions every year. The clay and bamboo sculptures are left to float away or sink, biodegrading to mix with silt, clay and the ashes of human remains, all eventually recycled as the chakra of life. (Author's emphasis).

CHAPTER 11 MUSIC OF THE VALLEY

OF THE FIVE SENSES, THE ONE WE ALMOST ALWAYS USE IS THAT OF SIGHT. The Valley Section is in essence a picture of an idealised terrain, with an outline, details and a space – usually blank – representing the sky. But a real valley has its river and the river makes a noise. And for some people that noise may be music to the ear.

The Geddes household was full of music. Anna Morton had studied Music for a year in Dresden, to a level at which she might, if she had to, teach music for a living. At the Summer Meetings she presided over social gatherings in the big flat in Ramsay Garden, bringing in such guests as Marjory Kennedy-Fraser (1857-1930), whose *Songs of the Hebrides* were a major contribution to the Celtic Revival and whom we now remember for her arrangements (and preservation) of Hebridean folk songs.

In the Ramsay Garden flat there was a pipe organ and in every house the family stayed in there was a piano or an organ. At home the day opened with a Latin hymn with Anna at the piano. Anna taught Norah and Alasdair sight-reading, singing by ear and had them compose tunes of their own. All the children were competent - or better – in a range of instruments. (For example, Arthur's contribution to maintaining morale during the Great War was to play his fiddle to refugees in the camps in France and, later, to cheer up airmen in camp and hospital in Lincolnshire. On Armistice Day he organised an ad hoc band and led them round the depot in celebration).

Patrick does not seem to have had any musical talent – perhaps a consequence of his Free Kirk childhood. But he did have a sentimental and intellectual attachment to the great Highland bagpipe and its 'very extraordinary scale in common use today'. He had John Duncan paint a series of murals in Ramsay Lodge illustrating the history and evolution of the bagpipe. No doubt his critics would say this was another manifestation of the 'failed sociologist'.

Geddes wrote to his older son Alasdair when the latter was aged twelve:

> You have always worn the costume of our Highland forbears, as I did as a boy, my father as a man, but we have both lost their language. Learn then this wider language, that by which the Celt at once appeals to every Scottish ear. Learn to play the old Celtic laments, the marches too, the pibrochs: blow loud and clear till men think of the long-lost Arthur returning in his might; croon too with the doves of peace; and chant like Columba and his brethren.
>
> In a summer or two you will lead some of our excursions, and almost from the first you will be able to start the march and help the fun.

This was no idle exhortation. In Paddy Kitchen's *A Most Unsettling Person* there is an awful photograph with the following caption:

> Children of Castlehill School, led by Alasdair, on their way to opening of playground in Castle Wynd, 1907.

But to the music of the Valley. The sound of water must have been one of the basic influences on the creation of music. Art music is full of the influence of water – storms, as in *The Flying Dutchman* of Wagner or as in Mendelssohn's *The Hebrides/ Fingal's Cave*, where we also hear that the seventh wave is supposedly the biggest. Mendelssohn also painted a picture in his *A Calm Sea and A Prosperous Voyage*. But Handel's *Water Music* was composed to be played on the water and is not descriptive of the Thames.

The most evocative music of the river must be Schubert's, in his song *Die Forellen* and his related *Trout Quintet*. Under the lilting top line there is a constant rippling and murmuring which exactly mirrors a lovely trout stream. In *Die Schöne Müllerin (The Lovely Miller's Daughter)* we hear the steady trundle of the millwheel, but is this enough? In Beethoven's *Pastoral Symphony* the second movement is 'By the Brook', a magical and extended scene of birdsong by the gently rippling brook, but again it is just an impression and not an overview.

Surely Strauss's greatest waltz *By the Beautiful Blue Danube* has something to offer? The words are not completely hopeless and – if the essence of the Valley Section is a progression – there are hints of progression in the text. The river rises in the Black Forest, flows eastward, and has castles, and mermaids (!) and sailors.

The Rhine is another great river with an abundance of maidens (Wagner), legends and stories. Schumann was peculiarly attached to the Rhine and set Heine's *Im Rhein*. His Third symphony is known as the *Rhenish* and is a mighty and colourful work, with a great deal of rushing movement. But it is more of a paeon to the great cathedral of Cologne and its medieval processions.

America, with its mighty rivers, should have something to offer. But Charles Ives' *Three Places in New England*, the last of which was *The Housatonic at Stockbridge* (completed in 1913) is disappointing. It is an atmospheric depiction of mists and running water and concludes with an almighty crash which signified nothing to me at first, but which now seems to me to be paying homage to the last few bars of Smetana's *Ma Vlast*. *Show Boat* (musical 1927, films 1936 and 1951, awards for Best Musical Revival 1991, and Best Revival of a Musical 1995) is concerned with racial issues – how does the mighty Mississippi stand on these? As sung by the great Paul Robeson as Joe, a black dock worker, *Ol' Man River:* '...doesn't seem to care what the world's troubles are, but 'jes' keeps rollin' along'.

Back in Europe, Peter Dawson (1882-1961), a fine, manly Australian baritone, was a great favourite on *Housewives Choice*, sometimes singing *Old Father Thames*. It is a good, healthy, optimistic, open-air song – but not very analytical!

> He never seems to worry
> Doesn't care for fortune's fame
> He never seems to hurry
> But he gets there just the same.

'Old Father Thames keeps rolling along' in every second verse, seeming to show that Raymond Wallace, the songwriter, had attended *Show Boat* once or twice! Topography does get a mention – from:

> High in the hills, down in the dales
> Happy and fancy free
> Old Father Thames keeps rolling along
> Down to the mighty sea.

Robert Burns seems to have been interested in his local rivers when he set new words to or rewrote the words of traditional Scots airs. But for Burns the river is a place where the poet can find the solitude in which to meditate on love and lost love, or, as in Tam o'Shanter, a locus for serio-comic events.

Clearly water in general and the river in particular have generated much sensitive and memorable music, but almost always the music portrays only one scene or one mood. The essence of the Valley Section is progression, from one valley type to another, and this is hard to find.

However, I can produce two quite divergent works which demonstrate progression.

Kenneth McKellar (1927-2010) was recently inducted into the Scottish Traditional Music Hall of Fame. This was:

> ...at last...recognition and rehabilitation for one of our country's greatest-ever singers, whose true legacy has long been clouded by a misleading and damaging 'tartan and shortbread' image.

Possessed of a fine tenor voice, McKellar moved from opera into a more popular domain where he was to provide the definitive interpretation of Burns' songs and modern arrangements of the likes of Marjory Kennedy-Fraser's *Songs of the Hebrides*. *Step We Gaily, Marching Through the Heathers* and *Hiking* songs are contemptuously dismissed, although very popular commercially, while *The Song of the Clyde* does not even get a mention.

Of all Scottish rivers the Clyde is the dearest to Kenneth McKellar - 'It flows from Leadhills all the way to the sea'. (The Miner is highest up Geddes's valley. Lead has been mined here, possibly in Roman times, but certainly from the 13th century). It 'meanders through meadows with sheep grazing there'. (And Geddes has his shepherd standing with his crook and a lamb). 'It borders the orchards of Lanark so fair'. (The small fruit and flower growers of Lanarkshire – the Peasants – are – alas! – no more. Put out of business by the state-subsidised North Sea gas of the Netherlands). By now we are down in the low ground, 'with towns on each side' where 'the hammers ding-dong is the song of the Clyde' and the river flows into the salt water of the Firth of Clyde. Till now the tune has been in a fine swinging 3/4 time. But now we switch to 6/8 and a patter song describing the leisure activities of what Glaswegians call 'doon the watter'.

It's a splendid song, with a real feeling of moving and rushing on from one activity to another.

And now for something (almost) completely different. And certainly more complicated. Let us start with the Elbe, which is said to be the longest river in Germany, although its headwaters are in the Czech Republic. It flows through the great cities of Dresden, Magdeburg and Hamburg before entering the North Sea. (The Rhein, Rhine, Rhin, Rijn passes through or gets its waters from Switzerland, Germany, Lichtenstein, France, Luxembourg and the Netherlands, so probably does not count as the longest German river!). What is now the Czech Republic was, in the nineteenth century, Bohemia, an important element of the Austro-Hungarian Empire. The official language was German, the capital was Prag and through Prag flowed the river Moldau.

All over Europe at that time there were restless minorities, eager to throw off what were seen as oppressive rulers and eager to rediscover their own cultures and even languages. Bedrich Smetana (1824-1884) took up the cause of his country's aspirations to independent statehood, to the extent that he is regarded as the father of Czech music. In the 1870s he wrote a series of six symphonic poems called *Ma vlast* (My country) one of which is *Vltava* – which concert promoters in Britain still persist in calling The Moldau! Poor Smetana! In 1874 he became completely deaf, had to give up all his public duties and died in a lunatic asylum in 1884.

Vltava was written in nineteen days and in it Smetana described the course of the great river, from the twinkling of two small springs through a series of landscapes, widening all the time, through Praha (Prague), rolling majestically until it joins the Labe (Elbe). It starts with a quietly sparkling theme realistically picturing the two little springs – the Cold and Warm Vltava – which come together into a single current which grows in volume and swings along. The music describes the river's course through woods and meadows. A farmer's wedding on its banks is described. Mermaids dance in the moonshine (not a component of Geddes's Valley Section!). On nearby rocks castles, palaces and ruins loom. The Vltava swirls into St John's Rapids, widening and flowing through Prague before majestically vanishing into the distance, ending at the Labe (or Elbe), with a satisfyingly *forte* chord.

Geddes would have loved the combination of nationalism and internationalism of Smetana's great work, while we can also admire how he tells the story of the river in such an absorbing fashion.

CHAPTER 12 ALTERNATIVES, ELABORATIONS AND VARIATIONS

GEDDES'S VALLEY SECTION IS A THINKING MACHINE, prepared for one purpose, to make one think. As we have seen with Figs 8 and 9, even as soon as 1910 Geddes was considering refinements and elaborations. Unsurprisingly, many others joined in, have done likewise and have taken the central idea further.

The writer was contemplating one version of the Valley Section when he had one of those light-bulb moments. 'But this is Von Thünen, this is *Die isolierte Stadt'*. Johann Heinrich von Thünen (1783-1850), a German landowner, in 1826 came up with a model of agricultural land use before industrialisation, which became very useful to economic history and Central Place Theory in geography.

Von Thünen postulated a city located centrally within an 'Isolated State', surrounded by wilderness. The land was completely flat, with no rivers or mountains (although later versions took account of these) and consistent soil quality and climate. There were no good roads, farmers taking their produce to the city by ox cart. Von Thünen also assumed that farmers behaved rationally to maximise profits. (Plate 6)

Taking into consideration the rent of land, labour input, production costs, market prices, distance to market and transport costs the result was four concentric rings. Nearest the city came dairy farming and market gardening. Before refrigeration these products had to be got to market quickly and their production was labour-intensive. Geddes's Section shows a spade next to the town. (Within the city of Edinburgh in 1914, although there was the Edinburgh and Dumfriesshire Dairy Company with a direct link with the south-west by the Caledonian Railway, there were still 64 'cow byres' within the city, with cowkeepers and irrigated meadows at Roseburn and Craigentinny. Geddes would have been well aware of this situation.)

The second ring looks surprising to us, because it was woodland. But it was the main source of fuel and building material at that period and costly to transport.

The third ring is of extensive field crops, such as grain, wheat, barley or rye, lighter to transport, less perishable than milk and requiring less labour input. (Geddes shows a rich peasant with a plough and a poor peasant with a mattock – further out.) The final ring is ranching. Animals require supervision but not constant attention. And they are self-transporting to the city for sale or butchering. And Geddes has a shepherd.

The wilderness is too far from the city for any type of agricultural product – and Geddes on the left has the miner, the woodman and the hunter. Having said this it is clear that there is more than a superficial resemblance between Von Thünen's Isolated State and Geddes's Valley Section. There is a difference of perception – Geddes's Valley Section is a view of a landscape, bringing in such factors as altitude, while von Thünen sees the world from above governed solely by economic factors. And – like a good German – he had a formula for measuring marginal productivity – R =Y (p – c) – Yfm!

Was Geddes aware of, or influenced by, von Thünen? So far, no evidence has been found, but one persists in searching.

FIG 13
Von Thünen meets the Valley Section
(Bulletin of Environmental Education)

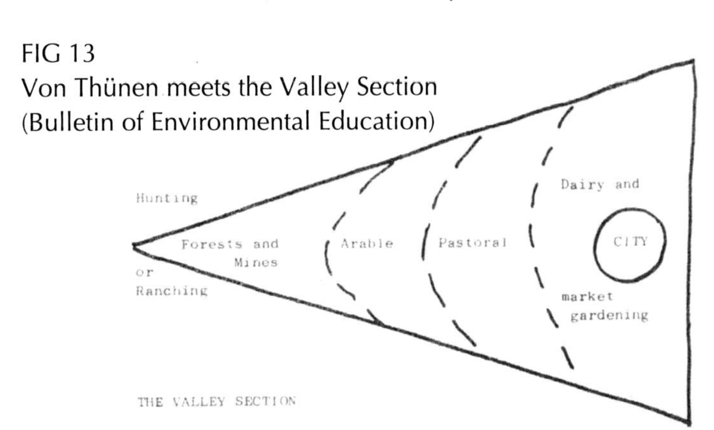

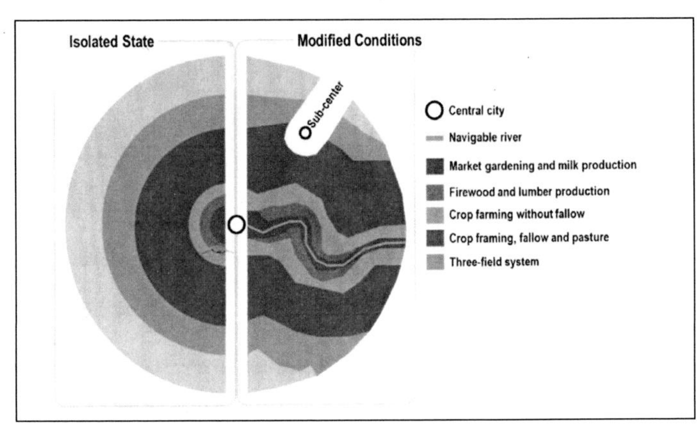

Rex Walford, a geographer of considerable repute, wrote a celebratory article about Geddes in the Bulletin of Environmental Education Vol 33, of January 1974. Almost as a throwaway conclusion, he threw in a rough sketch (very Geddesian!) of the Valley Section expressed in von Thünen's terms. (Fig 13)

Often described as a Peace Warrior, Geddes had a philosophical basis for his pacifism. He had doubts about 'survival of the fittest' and 'Nature red in tooth and claw'. One of his students said that he:

> ...traced the basal feature of all life to be the sacrifice of the mother for her offspring...saying...'So life is not really a gladiator's show: it is rather – a vast mothers' meeting'.

Geddes saw the First World War as the logical outcome of 19th century Darwinism. British and German minds were dazzled by the 'impressive nature-myth' of tooth-and-claw competition. And from 1914 the Prussian statecraft of brute force had openly been hurling this Darwinism of 'might is right' back at the Allies. During the First World War Geddes used his Summer Meetings to organise his thoughts on 'Wardom' and 'Peacedom', and in 1917 allied himself with Victor Bransford (1863-1930) to do something about it.

Bransford studied under Geddes at Edinburgh, was involved in the Outlook Tower and in 1903, with Geddes, set up the Sociological Society as part of the drift to London as the centre of activity.

In 1917, when the war was at its bloodiest, Geddes and Bransford set up a scheme of dazzling optimism – 'The Making of the Future'. Between 1917 and 1925 there appeared *The Coming Polity* (1917), *Ideas at War* (1917) *and Our Social Inheritance* (1919) and more than a score of pamphlets – *Papers for the Present*. They were startlingly revolutionary and hopelessly idealistic, but Geddes appeared to think that it was enough to point out the real causes of war and a logic for reconstruction for intelligent people to make a peace which would last.

But who would pay any attention to an ancient professor from an obscure Scottish university? As we now know, Clemenceau ('the Tiger') bullied the Peace Conferences into exacting revenge for the punitive peace of 1871 and making Germany pay the full costs of the war, thus creating the conditions for the rise of a strong man who would, in turn, exact revenge for the punitive peace of 1919.

A second edition of *The Coming Polity* appeared in 1919. This was illustrated and included three examples of the Valley Section, one of them (Plate 7) highly political.

FIG 14
The Nature Occupations (The Coming Polity)

Fig 14 is quite simple and straightforward, purporting to show 'the Nature Occupations'. The upper half is a slightly romanticised and accessible version of the Valley Section, while the lower half is a kind of gallery in which Geddes's seven occupations are represented in their appropriate clothing. In the text of the book there is the usual explanation for the occupations, but nothing to explain why they are depicted in this way.

FIG 15 The Valley Section and The Valley Plan
(The Coming Polity)

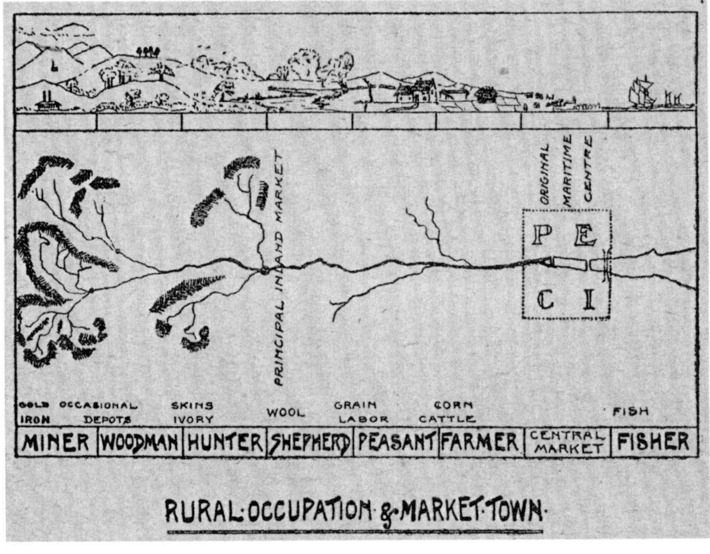

Fig 15 is interesting in that it combines the Valley Section with the Valley Plan. The upper part is a detailed sketch on which mine, forest, arable etc are easily identified, but where words would be expected there is only a row of eight empty labels.

The lower part is a plan of the river basin showing Rural Occupation and a Market Town. Along the bottom the eight boxes are for the usual seven types, plus a Central Market. Around the upper part of the river basin are mines and forests, while, where they come together, is a principal inland market.

Where the river becomes an estuary is the original maritime centre. Along the bottom are the products which each occupation would have fed into the markets – gold and iron, occasional depots, skins and ivory, wool, grain and labor (sic), corn and cattle, and fish.

Plate 7 is really quite simple. It shows two almost identical Valley Sections – but there are significant differences. The upper section is about Leadership. Leadership for what? For War. The lower section is about the softer Co-ordination. Co-ordination for what? The much gentler Civilisation. Who is to be our leader? The Hunter-Warrior? Or the Peasant-Shepherd?

Eight 'types' are listed beneath each, the seven usual ones plus the dwellers in the market town. In the upper section the dwellers are Townsmen, implying merely residence. The people in the lower section are, however, Citizens, implying that they work together for the common good. Time and time again, in his lectures and writing, Geddes emphasises the destructiveness and exploitation of the Miner, the Woodman and the Hunter while he celebrates the patience and industry of the Peasant. As for the Shepherd, he reminds us that the three great monotheistic religions – Christianity, Islam and Judaism – all stem from the lonely and caring shepherd in the desert night. And I am sure he would be familiar with the great works of art showing Christ's Epiphany, in which the Christ-child is revered by the Three Kings (one of them, Balthasar, black) representing temporal power. And who represent the common people? The shepherds, of course. And, perhaps labouring the point, let us remember that, in Riddle's Court in the Lawnmarket, beside the stained glass Valley Section Geddes had had created, is a companion piece of the Good Shepherd.

What is Geddes's forecast for society? In the upper option, on the right, we see cirrus clouds and an approaching depression, while, below, the sun is rising in a cloudless sky.

At the beginning, with the Geddes stained glass panel, we saw the two fighting eagles, reminding us that we had a choice.
He offered us an alternative – 'kakatopia' or the bad place of the future. Well over a century ago he knew we had a choice between ruining the planet and enabling it to sustain itself. In The Coming Polity Geddes makes no pretence of objectivity, he puts his cards on the table

In a few lines Geddes sets out a frame for the future.

Unfortunately, how many persons of influence did this reach?

Regional Survey was a major type of approach in Geddes's approach to the Sympathy element in his planning model of Sympathy, Synthesis and Synergy. Given that fact, one would expect the Valley Section to get a mention in Fagg CC and Hutchings GE, *An Introduction to Regional Surveying*, (Cambridge University Press, Cambridge,1930). Indeed, they give the Section a great deal of attention.

CC Fagg was a student under Geddes and a respected scholar of psychology while Hutchings had a life-time's experience in teaching fieldwork, especially in south-east England. Their *Introduction* was a valued standard work until, at least, the 1960s. Curiously, in the Patrick Geddes Archive at Strathclyde, GB249 T-GED/1/5/7 is a two-page paper, dated 1910, entitled 'An Introduction to Regional Surveying'.

FIG 16
Section through valleys (An Introduction to Regional Planning)

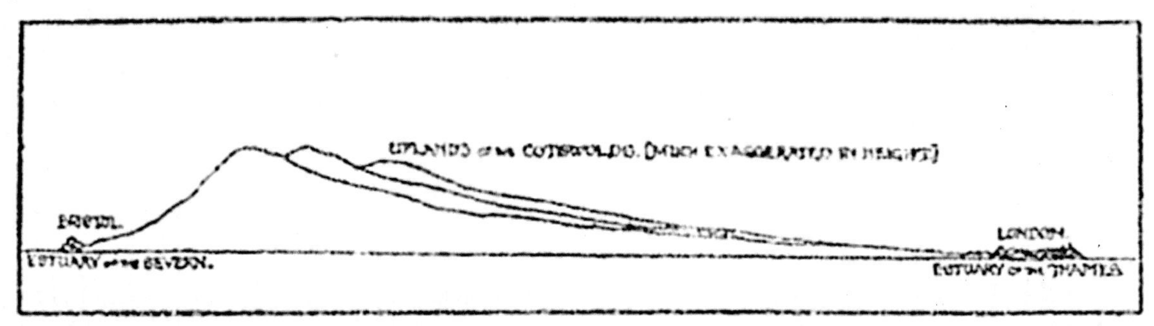

Fig 16 is not a Geddesian Valley Section – but has his three hills ('uplands of the Cotswolds') on the west. Fagg and Hutchings point out the universality of such a diagram – 'the Rhine-Danube valleys...are just an enlarged version of the Avon-Thames valleys. 'Such valley sections' they say, 'are the elemental diagrams of regional survey'.

FIG 17
The Zones of the Valley Section (An Introduction to Regional Planning)

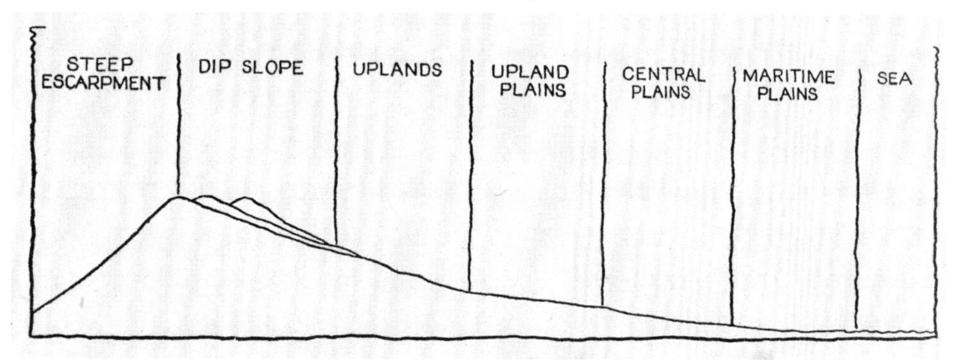

In Fig 17 they analyse the landscape in a totally different fashion. The human element is cut out, there are no rural types, only typical structural elements. Nor is there any suggestion that these elements determine the distribution of rural types.

FIG 18
The Valley Section, the Wandle Basin (An Introduction to Regional Planning)

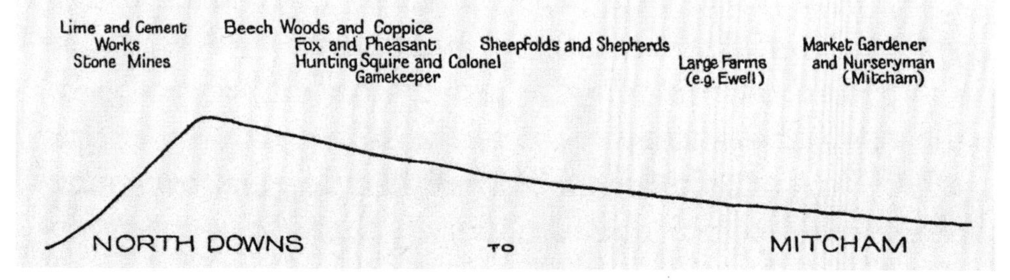

The River Wandle is an almost anonymous stream – barely shown and not even named on the Ordnance Survey Super Scale Atlas of Great Britain – near Croydon. Yet:

> … the small valley section of the Wandle basin presents an almost complete epitome of all the types and in their correct sequence on the section.

With the greatest of ingenuity and as a result of detailed regional survey, the authors have produced a convincing example of the Section in the real world, in contrast to Geddes's customary generalisations. This section also adds the dimension of time to the equation. In the ninety years since the book was published urban or suburban growth has simply wiped out some of Geddes's rural types. Lime works and stone mines on the Downs – taboo. Beech woods much diminished, coppicing obsolete. Sheepfolds, large farms, even market gardens and nurseries subdivided and built over. Geddes himself pointed out that his model only applied to certain environments and was subject to change.

FIG 19
The Valley Section with rustic types
(An Introduction to Regional Surveying)

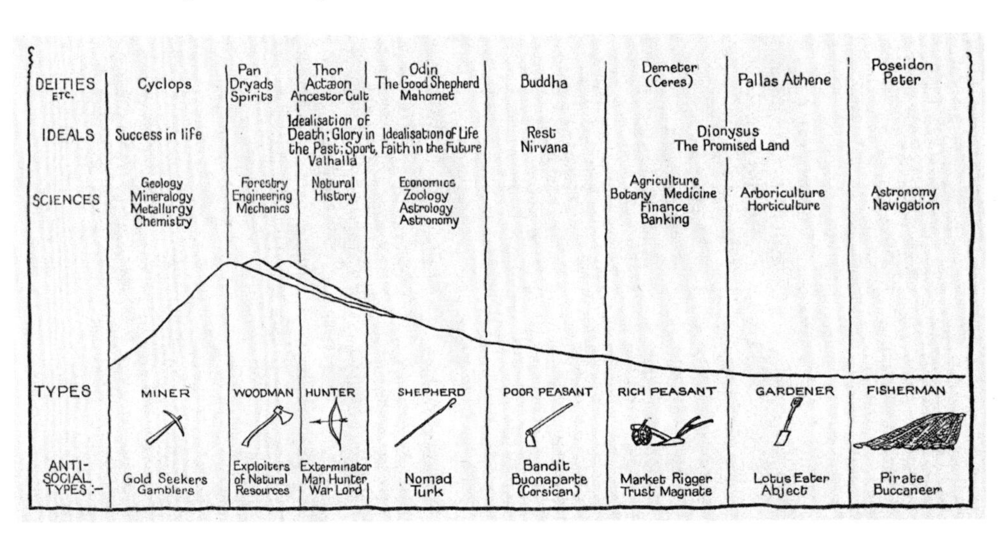

Fagg and Hutchings must have enjoyed drawing up their 'Valley Section with rustic types'. They have eight, giving them the opportunity to elaborate the differences between the poor peasant, the rich peasant and the gardener, and, of course, Geddes deconstructed these at length in his *Survey Graphic* talks and in *The Coming Polity*.

Now they consider Anti-social Types and seem to derive great enjoyment from this.

Geddes offered us an alternative – 'kakatopia' or the bad place of the future. Well over a century ago he knew we had a choice between ruining the planet and enabling it to sustain itself. And in *The Coming Polity* he made no pretence of objectivity, he put his cards on the table, and Fagg and Hutchings follow suit.

Thus the miner evolves into the gold seeker and the gambler, and, as we can see in Brazil today, the woodman becomes an exploiter. The hunter is an anti-social type par excellence while patriotic Frenchmen will be outraged to see their hero lumped in with the bandit under his Corsican name, and not classed as a man hunter or exterminator. We can see how the rich peasant can evolve into a market rigger or trust magnate but the gardener is more problematical. Do they mean that the gardener creates a private little world in which he idles his time away or are they suggesting that intensive cultivators live lives in abject poverty? Honest, hard-working fisherman to outrageous pirate and buccaneer – examples from history are many.

Linking the Sciences with the types is quite logical and acceptable, the only caveat being that Economics must surely underly all the types. The market rigger must be as au fait with the world of economics as the nomad.

Ideals and Deities, in a sense, go together, in the sense that deities are usually the understandable personification of ideals. Here, again, the associations are obvious. For example, the miner and the gold seeker aim to use the sciences to attain success in life. Contrast the idealisation of death, glory in the past and the ancestor cult of the hunter with the idealisation of life, faith in the future, Christianity and Islam of the shepherd. The Buddha, rest and Nirvana form a convincing existence for the poor peasant, while the pagan gods and the promised land offer a more nuanced lifestyle than mere success. The fisherman, it seems, has no ideals, being too busy merely surviving. Poseidon is not an attractive role model but we must remember that Peter was a fisherman and that Christ made those around him fishers of men.

The purpose of a model like the Valley Section is to stimulate thought and Fagg and Hutchings, in a younger generation, took Geddes's ideas and developed them further.

Arthur Geddes (1895-1968) was a 'brilliant and promising child – with nerves'. As the youngest in a very busy family he spent much of his early life on his own, although a Geddesian home education produced a good musician who was good with his hands and went on to become a distinguished lecturer in Geography at the University of Edinburgh. In the First World War he served with the Friends' War Victims' Relief Committee in refugee camps in France. With the introduction of conscription in 1916 he appealed – unsuccessfully - on conscientious grounds, only to be excused military service because he was declared medically unfit. The Friends were unable to take him back while his health was not satisfactory, yet he ended the war as 3rd Clerk in the Lincoln Stores District Park. He had a number – AG2832486RAF – a uniform, 'with zero to do'.

In the Twenties he spent most of his time in Montpellier, at his father's Collège des Écossais, trying to balance work on his doctoral thesis with the incessant demands of his ageing and occasionally irascible parent.

FIG 20
Arthur's concrete Valley Section (Marion Geddes)

Fig 20 is, quite literally, a concrete and naturalistic version of the Valley Section, with real, lumpy, hills, woodlands, terraces, a town, the sea with a boat – and a rising, optimistic, sun. On the frame, very faint, are the usual symbols. The label – E Languedoc (S. France) Montpellier Region – Cevennes to sea – shows that this is a real transect and not just a theoretical construct. On the other hand, concrete is obviously a much less effective means of conveying detail than ink on paper. Arthur's section is more ornamental than informative.

On the right and on the top border vague pencil marks show that this photo was a working document for Arthur.

CHAPTER 13 EXPLORATION

MABEL BARKER (1885-1961) WAS GEDDES'S GOD-DAUGHTER and – to use her own words – '... a troublesome assistant who will not be dismissed'. A bit of a tomboy and an expert climber, she plunged vigorously into Geddes's world, especially his commitment to regional survey and peace initiatives. Her career saw a continuing search for a post which would match her ideals. At the same time she was involved in the whole range of Geddes activities, from organising Peace conferences to writing labels for *The Patrick Geddes Collection of Edinburgh Photographs*.

During the First World War Rudmose Brown of Sheffield University was requisitioned by the Admiralty to write handbooks on Siberia, Finland, Norway and Sweden. Mabel took over his classes and, in so doing, became the first woman to teach Geography in a British university. The Netherlands were neutral and were full of destitute Belgian refugees and internees. At Elizabethsdorp, a camp for wives, children and elderly relatives of Belgian soldiers interned close by, under the aegis of the Friends' War Victims' Relief Committee, she organized paid employment, evening classes and programmes of outdoor activities. After the war she taught at a progressive school till 1925, when she joined Geddes at his Scots College in Montpellier. Her last venture was to start her own school in Cumbria along Geddesian lines.

The Le Play Society (and the Institute of Sociology) was founded by Geddes in 1930 for '...the development of the theory and practice of geographical and sociological fieldwork'. The history of these two organisations is unbelievably complex and tendentious, but all we need to know at the moment is that Le Play House was committed to Regional Survey and foreign field work. Between 1921 and 1952 85 'or so' trips were taken to a diversity of European locations.

In August 1926 a Le Play House Student Tour to the French Pyrenees was directed by Professor W Stanley Lewis and led by DT Williams. Their paper of 16 pages - *THE UPPER ADOUR; A Study of a Pyrenean Valley* – was published in The Sociological Review. The study area was 'a typical example of the prevalent transverse Pyrenean Valley' above the spa and ski resort of Bagnères-de-Bigorre, 21 Km southeast from Tarbes.

It is a systematic study of a small region. The headings can be deduced - physical features, geology, vegetation zones, climate, soils and agriculture, 'folk-type' and 'folk-life', social evolution etc. There are three illustrations, a locational map with south (and Spain) at the top (!), a section **across** the Adour valley and the plan of a peasant house (12 feet by 10 feet).
When they come to industrial activities, Lewis and Williams record that all the industries have local associations and classify them under the headings of Mining, Forestry, Animals and Human. Suddenly they write:

> If these be considered in conjunction with the pastoral, agricultural and gardening pursuits, a close approximation to the valley-section (N.B. no capitals) may be traced. In close relation to the whole there exists a varied manifestation of the professions, responses to the more intricate material and to the spiritual demands of the community.

Then they give detail. 'Mining'= two marble-working factories, slate-quarrying and clay working. Forestry feeds three sawmills and joineries and is a good example of industrial inertia. There was cheap locally-generated electric power, cheap (Spanish) labour and a good local market (a railway wagon works employing 350). But:

> The commune of Bagnères...has little reserve of timber for local industrial uses. It is obvious that great reliance must be placed upon foreign supplies of timber.

A useful reminder that the Valley Section is not set in stone but can evolve as social and economic factors also evolve.

After examining so many abstract models and theoretical variations of the Valley Section it is refreshing to come across a 'worked example', a sign that the concept certainly has – or had – validity.

Lewis and Williams' paper shows that the study group followed a systematic procedure which may, or may not, have been laid down on paper. In 1939 Mabel was responsible for producing for the LePlay Society *Exploration: Regional Survey: Get to know your own Place and Work and Folk*. It starts: 'Walk all over your region till you know it well enough to draw a sketch map from memory...'.

FIG 21
The basic Section with simple advice (*Exploration*)

Basically, it is a little (16 pages) handbook which sets out methodically how to carry out a regional survey. The very first page (after the title page) is headed by the simple version of the Valley Section and a list of the tools needed by an explorer. In other words – Be Prepared – the motto of the Boy Scouts. It goes on to set out 'tasks' from General Approach to Camping and Rambling. It was republished by the Patrick Geddes Memorial Trust in 2007 with an Introduction by Kenneth MacLean and Walter Stephen.

Around 1946 the writer acquired a copy of *The Boy Scout's Handbook*, 'produced to Wartime Economy Standards'. It carried details of all the Scout tests and badges and how they could be earned. One of the higher level badges, for which only First Class Scouts were eligible, was the Explorer's badge. (The high entry level ensured that those entering for the badge were reasonably able to handle themselves when surveying alone in the field). Fifty years later, when I put the two together, I was surprised to find how much of the *Handbook* advice was an exact copy of Mabel's publication.

One of the questions we will never be able to answer is how many young people were unconsciously introduced to Geddes and his ideas through the Scouting movement.

CHAPTER 14 *THE LATE NORMAN THOMSON*

AS A SUPPOSEDLY MATURE STUDENT moving towards a new career, I spent three weeks in my old school, much of it in the presence of Norman Thomson. He claimed to remember me from playing rugby against me. I think he was mistaken. He never played in any first team. He was too nice. He was a gentle giant.

A few years later we met again, when he was Principal Teacher of Geography at the brand-new Broughton High School in Edinburgh – in a superb department influenced by Her Majesty's District Inspector of Schools, Dr Gilbert (a geographer). Norman went on to become a Lecturer in Geography at Moray House College of Education, finishing off as Head of the Social Studies department.

At a time when the appeal of Geography in schools was being stifled in a cold climate of quantification, dreary concepts and the emergence of 'homo-tickieboxus' Norman was vigorous, with the Scottish Association of Geography Teachers, in bringing back into the classroom colour and humanity. With Kenneth Maclean, PT in another excellent 'Gilbert department' at Craigmount in Edinburgh, then Lecturer in Geography at Moray House College of Education, who returned to teaching as PT at Perth Academy, he wrote *S1-S2 Geography*, published in 2003.

Chapter 1 is 'Introducing Place, People and Work'. Sir Patrick Geddes comes in in the second paragraph and there are photographs of Geddes, his family house, the Outlook Tower and renovated housing in Edinburgh – with assignments. Overall, this is an attractive and exciting textbook, fitting into the 5-14 curriculum. While no school was obliged to use it, it must have been used by thousands of young Scots. It could be said that more people have been introduced to Geddes through Maclean and Thomson's pages than from all the learned writings combined!

Norman had a wry sense of humour and this was demonstrated in two Valley Sections he prepared for the Scottish Association of Geography Teachers. NLS 10619 is a folio of Geddes's notes in the National Library of Scotland and from it, with a great deal of imagination, Norman created 'Patrick Geddes: The Teaching of Geography' (Fig 22) and 'Patrick Geddes: The Valley Section and Teaching Maths' (Fig 23).

FIG 22

Patrick Geddes: The Teaching of Geography (*Scottish Geographical Education: Teachers, Texts and Trends*)

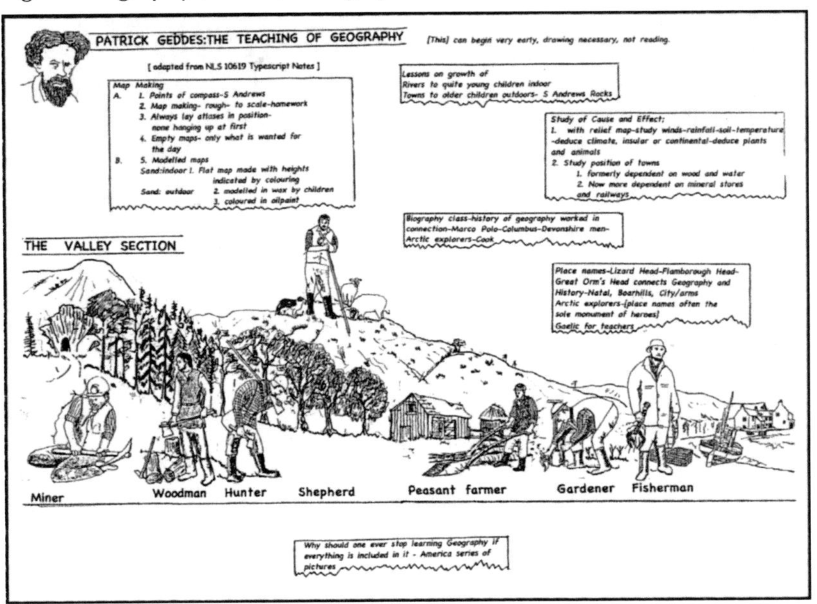

The lower half of Fig 22 is a free sketch of the seven rural types in 20th century dress and with 20th century equipment. For example, most of them wear wellington boots and safety helmets. Under the diagram appears the following:

> Why should one ever stop learning Geography if everything is included in it?

The upper half states that (The teaching of geography) 'can begin very early, drawing necessary, not reading'. What appear to be scraps from a notebook give suggestions on – Map Making, Lessons on growth, Study of Cause and Effect, Biography class-history, Place-names.

In one sketch Norman Thomson outlined a curriculum which could support the Geddesian philosophy of education of *Vivendo discimus* ('By living we learn').

FIG 23
Patrick Geddes: The Valley Section and Teaching Maths (*Scottish Geographical Education: Teachers, Texts and Trends*)

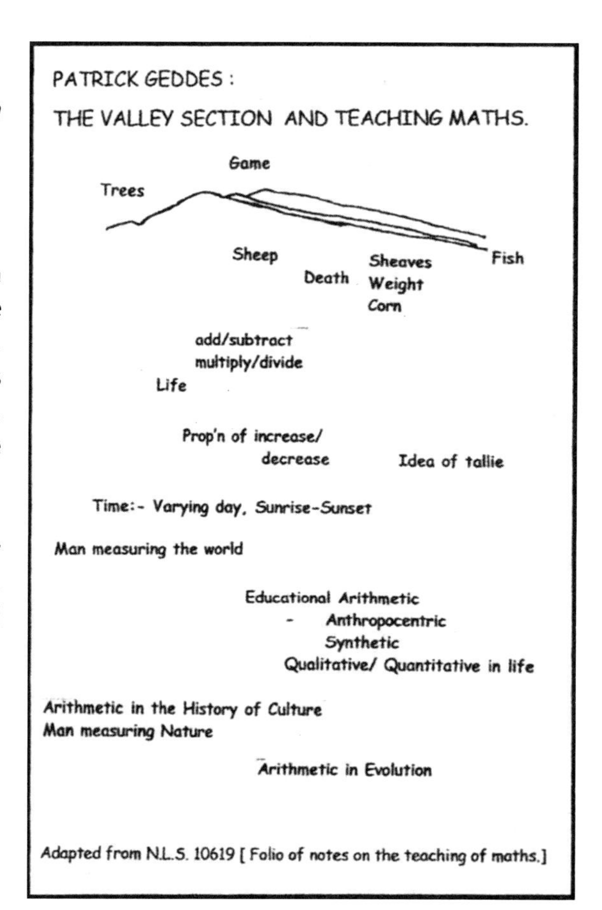

Geddes was good at Maths in school and worked for eighteen months in the bank, presumably mastering Arithmetic. So his notes must have some interest. In Fig 23 we see Norman Thomson indulging in a little *jeu d'esprit*, the sort of thing one might do in the staffroom once the crossword has been done. His valley profile still has three hills but has no occupations, only trees, game, sheep and fish. There are lots of things that could be done. But where? And when?

One can see his topics being used as illustrations or examples at particular points in the Maths curriculum but I do think Geddes was trying too hard. Consider the dangers of the whole class counting sheep in an overheated classroom!

CHAPTER 15 THE VALLEY SECTION TODAY

SURELY THE VALLEY SECTION IS OUT-OF-DATE? Or, at best, only of historical interest? In Scotland a minute quantity of gold comes from the Highlands but our real mineral wealth is under the sea. Greater mobility in our times means that people no longer have to live beside their work, or work within walking or cycling distance of their homes.

Consider the Woodman. In the wake of the Second World War there was a terrible shortage of timber in Britain, while timber and wood products were our second biggest import by value. The Forestry Commission began an enormous planting campaign, which had the secondary aim of bringing an active population into depopulated areas. Little clusters of forestry houses were built in the Highlands, and a village complete with shops and a school at Ae in Dumfriesshire. In the 1970s I was arranging some educational work at Glentress, just east of Peebles. I found that, of the 35 people who worked there, only the manager lived in Peebles. The remainder commuted from Edinburgh – and not all in one bus!

As the writer drives south from the centre of Edinburgh – from the Central Business District – he notes a ring of transition to working class housing, then of residential housing, and then another ring of transition, the rural-urban fringe or the Green Belt and recalls Burgess's Concentric Zone Model of 1926. Next day he drives out westward from the CBD, through Gorgie to Stenhouse and Sighthill, a sector of mainly working class housing with industry, railway and canal. This time he recalls Homer Hoyt's Sector Model of 1939. Which model is 'right'? And which 'wrong'?

Two models, each with some validity, but: 'As with all simple models of such complex phenomena' their 'validity is limited'. But they provide plenty opportunity for debate.

In Chapter 12 we came across Von Thünen's Isolated City and its rings of similar activity based on rational factors. Clearly his model is way out of date. We import wood chip from across the Atlantic instead of growing our fuel near the city limits, for example. But his central idea, that economic activity sorts itself out according to the intensity and nature of its inputs, still seems to hold good. For example, William Alonso's *Location and Land Use*: *Toward a General Theory of Land Rent* of 1964 built upon the Thünen model to account for intra-urban variations in land use.

Coming back to Geddes and his Valley Section, he was never dogmatic or universal about it. In fact, he could be quite diffident. He acknowledged his debt to LePlay, saying that: 'Our interpretation of LePlay and his method has indeed been implicit throughout much of' the Survey Graphic lectures. In fact he points out that Lecture II is about Place, Lecture III is about Work and the later lectures are about Folk. For Geddes *'Lieu, Travail, Famille'* were the 'open sesame' for studying the enduring factors and civilisation values of any region or city - 'a foil to the historians and the heralds'

But not any region. Although he draws morals and examples from every continent and every society, he makes it plain that his Valley Section really applies only to those environments he knows well – Scotland, Western Europe, the Mediterranean – although there is no reason why we should not, for example, see how well the Valley Section fits a section from the Andes to the Pacific.

In discussion about the Sixpenny Valley Section, of Chapter 9, a double section, with the essential occupations, their names and their 'successors' below, Geddes realises that:

> These are not all represented in every valley; and any, of course, may occupy it in very different proportions. Still this outline serves all the better for general purposes.

He might also have referred to changes over time (as he does in the lectures) and would not have worried that his original construct was no longer valid.

For Geddes, his Valley Section was not a finished product, sacred and beyond improvement. It was a model, a stimulant and a target for constructive thinking about our world.

Just as he said that he was 'the little boy who rings the bell and runs away' his Valley Section is the grit in the shell of the oyster that produces a pearl – in his case creative thinking about how the world works.

Here are three examples of the way in which the Valley Section lives on.

Geddes would have approved of Juliet Robertson. Her *Dirty Teaching: A Beginner's Guide to Learning Outdoors* was published in 2014 and has been reprinted several times since. The 'book is for primary teachers and student teachers who want to teach outside'. She has the good sense to acknowledge the support and advice given along the way by (among others) Dr Walter M Stephen and the Sir Patrick Geddes Memorial Trust.

In her introduction she suggests:

> The idea of using *place* as a key part in the learning process comes from the work of Sir Patrick Geddes (1854-1932), a Scottish town planner, biologist and educator, known for his progressive views, who developed the concept of 'think global, act local'. He also advocated a 'hands, heart, head' approach to learning.

Dirty Teaching is essentially a handbook systematically setting out why and how we should teach outdoors and is full of illustrations, examples and ideas. Chapter 12 is 'Embedding Outdoor Learning' and Idea 12.5 is 'Use the Valley Section as an evaluative activity'. She finds the Valley Section particularly helpful both as a class teacher and when looking at outdoor learning at a whole-school level. She gives examples of how subjects, like maths, can be made relevant to real life.

Her conclusion is that:

> The Valley Section provides us with an open-ended way of thinking about how we teach. It can form the basis for reflecting more deeply about embedding outdoor and place-based approaches that goes beyond a tick-box mentality.

Robertson's *The Valley Section with basic occupations* (Plate 8) is an adaptation of Mabel Barker's in *Exploration* (Fig 21) but demonstrates very well that the Valley Section still has a meaning in the 21st century.

There are still seven occupations, but they have changed since the Valley Section first appeared in the early 20th century. The Miner has gone, to be replaced by the Geologist. *Exploration's* Victorian pithead has been modernised and now shows the pitshaft. The Woodman is now a Forester, implying control, and has conifers in rows. The Hunter has vanished, to be replaced by a Conservationist. His symbol is a dove carrying an olive branch. (This resonates with the portrayal by Geddes and Fagg and

Hutchings of the Hunter as a negative force in society). The image is of deciduous trees at different stages of growth, with a white fence giving controlled access.

The Shepherd and the Peasant have both disappeared, to be replaced by a Farmer, with his tractor trundling through big fields, and a Gardener (tools – trowel and hand grubber). The Gardener has a mass of flowers, but whether these represent a zone of market gardening (as with von Thünen) or suburban housing is a matter for discussion. The Farmer has disappeared, a victim of urban sprawl, and is replaced by the Town Planner, with his compasses and multi-storey buildings. The Fisher is the only occupation which has not changed, although his technology has changed.

Juliet Robertson is to be congratulated for breathing new life into what some might regard as an outmoded notion.

There are two Japans. There is the Japan of Hokusai and Mount Fuji, of cherry blossom and the haiku, of Gilbert and Sullivan and *The Mikado*. Then there is the bustling industrial giant, a leader in technology, the ultimate in urbanisation awash with liberalism and Marxism. In 1906 Geddes must have been very conscious of these two Japans. One can guess where his sympathies lay, but there is no sign of his interest. Nor was there any sign that Japan was interested in Geddes. At one time it was thought that Japanese interest in Geddes grew out of the need to rebuild Tokyo after a huge earthquake in 1923 but, in fact, reconstruction was planned by young Japanese engineers using the German system of planning.

Mike Small, in his essay on *Universalism and the Genus Loci* (2007), quotes Professor Toshihiko Andoh:

> Generally speaking, those who got interests in Geddes have been the intellectuals who searched for another way of modernisation or urbanisation. For example, Nakagawa, the Secretary of Home Office who wrote a letter to Geddes in 1909, was a member of a Home Office research team on the garden city movement which searched for a more balanced way of urbanisation for the local governments. Perhaps he found Geddes's *City Development* during that research?

Professor Sadakata, a geographer involved with Yamaguchi, recorded:

> Michitoshi Odauchi, human geographer out of office, introduced the idea of Patrick Geddes and applied it for his

> regional studies around 1930. I think he is the first Japanese who referred to Patrick Geddes in detail. But he did not develop the Patrick Geddes idea further as German ideas began to have a stronger influence.

Nevertheless, there is a small outpost of Geddesian ideas in the south-west of Honshu, the main island of Japan. Yamaguchi. Yamaguchi (City) has a population of 198,971 – the size of Aberdeen – and has one university and the Yamaguchi Center (sic) for Arts and Media (YCAM). It is a large modern building 'defined by the pursuit of new artistic expression incorporating media technology'. The centre hosts exhibitions, performances, screenings, workshops and residencies.

In 1995 was taken: 'a first great step to knot a perpetual friendly ties (!) between Edinburgh and Yamaguchi'. At least three Edinburgh Geddesians were involved, Mike Small (above), Murdo Macdonald, a PGMT Trustee and Professor Duncan Macmillan. Murdo's subject was *The Visual Thinker: Patrick Geddes* and included a substantial description and analysis of the stained glass version of the Valley Section. Duncan Macmillan's contribution was 'By leaves we live' - an aphorism of all ecology.

This is a really interesting relationship, as it combines Geddes's biology-based ideas with the cutting edge of modern technology. But, as Mike Small writes:

> It would be easy to overplay the importance of Geddes in Japan...but what is encouraging is that we now have what one contributor in Yamaguchi called 'a thousand Geddeses' in countries across the world. We will need them.

Plate 9 is from the cover of the 2018 *Yamaguchi Valley Section* – Reflections event. (By 2018 YCAM had become YICA – Yamaguchi Institute of Contemporary Arts – and was celebrating its 20th anniversary). No realistic Valley Section here, merely the symbolic attributes of the Geddes rural types.

Plate 10 appears on page 60 of *Chapelton: The Making of a Town* with the caption 'The neighbourhood structure'. The caption continues:

> Chapelton is arranged into seven walkable neighbourhoods including a town centre, each comprising everyday amenities and residential streets of varying designs and densities.

Fig 21 is the straightforward, simple Valley Section. What on earth has this to do with the grandiose Chapelton development?

Chapelton of Elsick, 16 kilometres south of Aberdeen, is one of the UK's biggest new settlements, with the long-term aim of up to 8,000 houses within seven neighbourhoods. The town is being built on land owned by David Carnegie, 4th Duke of Fife (who is also the director of the Elsick Development Company which is overseeing the development) and four local farming families. The master plan was developed in consultation with over 5,000 local people and professionals, local and national.

Behind such a large and complex project there must be some presiding genius. For Chapelton this was Andrès Duany, an American architect, an urban planner, and a founder of the Congress for the New Urbanism, an international urban planning movement intended to offer an alternative to suburban sprawl and urban disinvestment. His company (DPZ) has completed designs and codes for over three hundred new towns, regional plans and inner-city revitalization projects. He has written several books and held various academic posts.

Duany found Geddes's ideas on the reactions between regions and their built environments, the interaction between urban, regional and ecological conditions – and the Valley Section – interesting. Duany's theoretical exploration of the transect evolved into his own model for understanding the progression of urban form from uninhabited countryside to city core. This became a tool for establishing 'a suitable graduation of intensity' within the built environment.

Geddes's Valley Section had seven types, DPZ's Smart Code divides the rural-to-urban transect into six environments, or zones. These are: rural preserve, rural reserve, sub-urban, general urban, urban centre and urban core. This transect methodology was brought to Scotland for Chapelton in 2010. The work on the ground is far from complete, but it is an interesting exercise for us to go back to Plate 10 and see if we can detect Duany's zones in and around the seven neighbourhoods.

Finishing our survey of the Valley Sections with Chapelton gives a comforting sense of closure. Within a few miles of Chapelton is a string of former fishing villages – Portlethen, Newtonhill, Muchalls, Cowie – and Stonehaven, once an important fishing port. On Thursday 7 August 1879 the Scottish Zoological Station ('the only building of its kind in Great Britain') was inaugurated, in Cowie. Behind the project was Professor James Cossar Ewart, Professor of Natural History at the University of Aberdeen. At the inauguration, on the wings, probably stood Geddes.

At University College London, Geddes had shared laboratory space with Cossar Ewart, then Conservator of Museums. Geddes was headhunted, sent to Roscoff in Brittany, then to the Sorbonne and then to Naples, the first zoological station (with aquarium) in Europe. Back in Scotland he wrote two articles for *The Scotsman* on the need for such a station in Scotland. A few weeks later, Cowie was opened and Geddes was appointed to run the establishment. When it closed for the winter Geddes went off to Mexico collecting for Kew, Huxley and Edinburgh. On his return he was appointed demonstrator of Botany at the University of Edinburgh (based in the Royal Botanic Garden) and lectured on Natural History in the School of Medicine.

The sense of closure is enhanced by the knowledge that Ben Tindall, excellent Scottish architect and planner, former Chairman of the Sir Patrick Memorial Trust, was involved as part of Andrés Duany's charette design team. Indeed, the very first layout was done by Tindall at breakfast!

There is a kind of beauty in the Valley Section story. At Cowie Geddes collected and classified, but also learned how to draw the lay public into a love of nature. In Mexico, temporary blindness forced him to develop 'thinking machines', the best known of which was the Valley Section. And then, after more than a century, appears, just a few miles away, a 'Brave New World' of urban planning and development.

Very satisfying.

Geddes 'did what he could'. He had his regeneration schemes and his little garden suburb, but they were always under-financed and under-managed. But his ideas were there and, wherever he is now, he must look down benevolently to see what is happening today in his name.

I am happy to let Rex Walford have the last word.

> No doubt quite rightly the Valley Section appears to the modern eye as picturesque as a William Morris wallpaper. Nevertheless it is still worth consideration for the insights it offers the environmentalist and the educator. Geddes was, and remains, one of the few who provide a vision of the human and natural world alternative to that of the analytical scientists who reduce everything to an explanation rather than an interpretation. Indeed, in this and many other ways, he expressed ideas whose reality is only now becoming accepted as we enter a new phase of environmental awareness.

Appendix

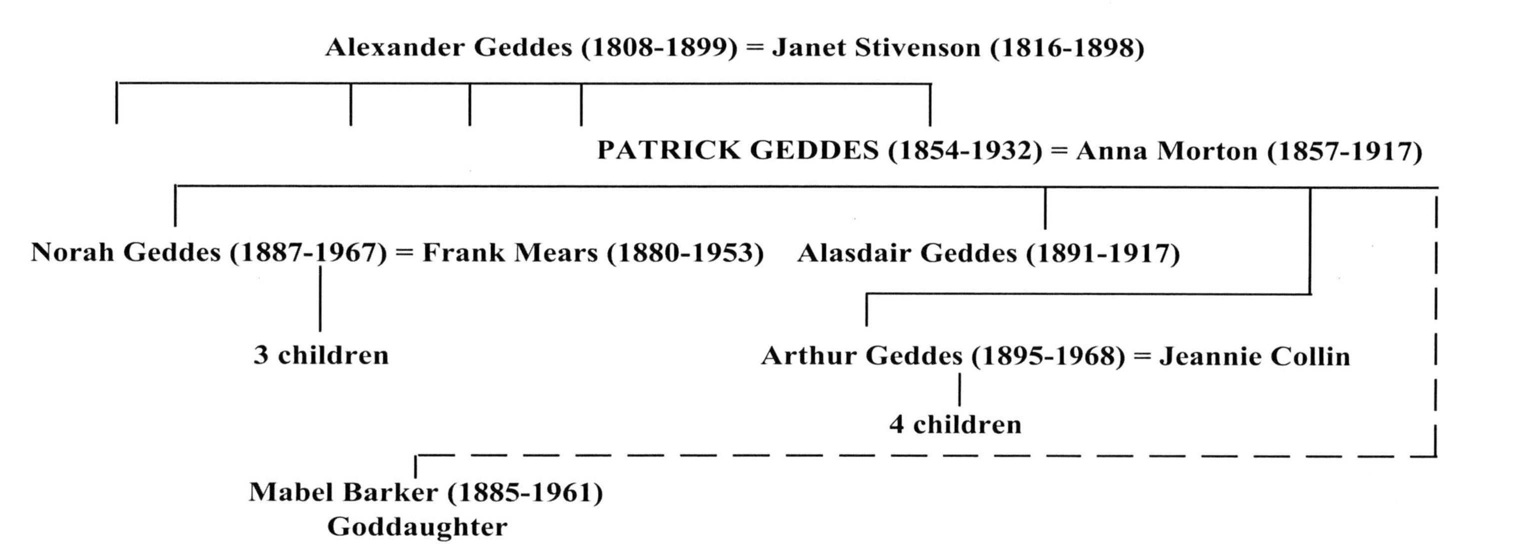
THE PATRICK GEDDES FAMILY (SIMPLIFIED)
Alexander Geddes (1808-1899) = Janet Stivenson (1816-1898)
PATRICK GEDDES (1854-1932) = Anna Morton (1857-1917)
Norah Geddes (1887-1967) = Frank Mears (1880-1953)
Alasdair Geddes (1891-1917)
3 children
Arthur Geddes (1895-1968) = Jeannie Collin
4 children
Mabel Barker (1885-1961)
Goddaughter

Bibliography

Bouche, Patrice, *Patrick Geddes's (e)utopian belvedere in Southern France*, (Planning Perspectives 29(1), June 2014)

Branford V and Geddes P, *The Coming Polity: A Study in Reconstruction*, New and revised Edition (Williams and Norgate, London, 1919)

Fagg CC and Hutchings GE, *An Introduction to Regional Surveying*, (Cambridge University Press, Cambridge,1930).

Geddes, Sir Patrick, *Collecting Cities: Images from Patrick Geddes*, (Cities and Town Planning Exhibition, Collins Gallery, Glasgow, 1999).

Macdonald, Murdo, *Patrick Geddes e L'Intelletto Democratico*, (in Spazio e Societa, vol 68, Milano, 1994).

Macdonald, Murdo, *The Visual Thinker: Patrick Geddes* (Edinburgh-Yamaguchi '95, Yamaguchi, 1995)

MacLean, Kenneth and Stephen, Walter (eds.), *Exploration: Get to know your own Place and Work and Folk* (Hills of Home, Edinburgh, 2007)

MacLean, Kenneth and Thomson, Norman, *Scottish Geographical Education: Teachers, Texts and Trends* (Scottish Association of Geography Teachers, 2007)

Ogilvie AG (ed), *Great Britain: Essays in Regional Geography*, (Cambridge University Press, Cambridge, 1930.)

Élisée Reclus, *Histoire d'Un Ruisseau* (Bibliothèque d'Èducation et de Récréation, Paris, 1881)

Roberts, Paul, *Chapelton: The Making of a Town* (Turnberry Consulting Ltd, 2018)

Robertson, Juliet, *By Leaves We Learn*, (Reforesting Scotland, Issue 58, Autumn/Winter 2018)

Robertson, Juliet, *Dirty Teaching: A Beginner's Guide to Learning Outdoors*, (Independent Thinking Press, Carmarthen, 2014.)

Stephen, Walter (ed.), *A Vigorous Institution: The Living Legacy of Patrick Geddes*, (Luath Press, Edinburgh, 2007.)

Stephen, Walter (ed.), *Learning from the Lasses: Women of the Patrick Geddes Circle*, (Luath Press, Edinburgh, 2014.)

Stephen, Walter, *On the Trail of Patrick Geddes*, (Luath Press, Edinburgh, 2020.)

Stephen, Walter (ed.), *Think Global, Act Local: The Life and Legacy of Patrick Geddes, New Edition*, (Luath Press, Edinburgh, 2014).

Stephen, Walter, *Where was Patrick Geddes born?* (Hills of Home, Edinburgh, 2008)

Thompson, Catharine Ward, *Geddes, Zoos and the Valley Section* (Landscape Review, v10, Lincoln, NZ, 2004)

Thompson, Catharine Ward, *Patrick Geddes and the Edinburgh Zoological Garden: Expressing Universal Processes Through Local Place* (Landscape Journal,v 25, University of Wisconsin, 2006)

Town and Country Planning Association (Great Britain), *Bulletin of Environmental Education* (London, 1977)

Walford, Rex, *Geography in British Schools 1850-2000* (Woburn Press, London, 2001)

Welter, Volker M, *Biopolis: Patrick Geddes and the city of life*, (Cambridge, Mass/London, c2002)

Welter, Volker M and Lawson, James, *The City after Patrick Geddes*, (Oxford, c2000)

Other Books from Patrick Geddes Memorial Trust

Think Global, Act Local

Walter Stephen

ISBN 978 1 910745 09 0 PBK £12.99

In turns a gardener, biologist, conservationist, social evolutionist and town planner, Patrick Geddes spent many years conserving and restoring Edinburgh's historic Royal Mile at a time when most decaying buildings were simply torn down. With these plans of renovation came the importance of education – as the development of the Outlook Tower, his numerous summer schools and his Collège des Écossais in Montpellier illustrate. It is in India where his name is most widely known. It was here that possibly the greatest example of Geddes' belief in 'people planning' can be seen and which took the form of pedestrian zones, student accommodation for women, and urban diversification projects in Edinburgh.

A Vigorous Institution

Walter Stephen

ISBN 978 1 905222 88 9 PBK £12.99

Patrick Geddes was a polymath. Gardener, biologist, conservationist, town planner, warrior for peace, social evolutionist; he achieved an incredible amount for one man. This book takes a new look at Geddes' life, drawing on newly discovered material to try to come to an understanding of Geddes' drive and success. How much of an anarchist was he? How influential were his home and childhood experiences? Why did he change his name and why – until the publication of this book – was his birth site shrouded in mystery?

Learning from the Lasses

Walter Stephen

ISBN 978 1 910021 06 4 PBK £12.99

It is well known that Patrick Geddes shaped the cityscape of Edinburgh, but for the first time Walter Stephen turns the lens onto the strong, wilful women who influenced the revolutionary man – and who were in turn influenced by him.

From his wife and mother in Scotland, to a nun in India and a Marchioness in Ireland, this insightful volume shows the wide range of women across the globe whose lives intertwined with Geddes's, whether professionally or personally. Delving deeper into Geddes's personal life than ever before, Walter Stephen and his fellow Modern Geddesians go beyond the surface of the Scotsman's acclaimed works to reveal the female characters that shaped him throughout his life.

The Interpreter's House Revisited

Stephen Hajducki and Walter Stephen

ISBN 978 095551906 2 PBK £6.99

Tells the story of the major financial crisis affecting Patrick Geddes's Outlook Tower, the steps taken to sort things out and the interesting characters who combined to keep the enterprise going.

The Interpreter's House Revisited

Stephen Hajducki, Walter Stephen with Tomas Slater